T0297577

Climate Change

Climate Change

Alternate Governance Policy for South Asia

Ranadhir Mukhopadhyay
S M Karisiddaiah
Julie Mukhopadhyay

Elsevier
Radarweg 29, PO Box 211, 1000 AE Amsterdam, Netherlands
The Boulevard, Langford Lane, Kidlington, Oxford OX5 1GB, United Kingdom
50 Hampshire Street, 5th Floor, Cambridge, MA 02139, United States

Library of Congress Cataloging-in-Publication Data
A catalog record for this book is available from the Library of Congress

British Library Cataloguing-in-Publication Data
A catalogue record for this book is available from the British Library

ISBN: 978-0-12-812164-1

Publisher: Candice Janco
Acquisition Editor: Laura S Kelleher
Editorial Project Manager: Tasha Frank
Production Project Manager: Omer Mukthar
Designer: Matthew Limbert

Typeset by Thomson Digital

This book is dedicated to those eternal shepherds who remain busy in integrating human with nature

Contents

Preface

Innovation is fundamental to the future growth. Society needs to change with time, and many of the beliefs and theories of the present day would be redundant in near future. Research is the key tool to cope with such an existential changeover and to meet the ever-increasing human aspiration. To make that happen one needs to broaden the skill, education, and human capacity, and make research meaningful to the society. It is also equally essential to enliven the progress made in the fields of science and technology with comparable intensity of art, ethics, values, and egalitarianism to enable sustainable, balanced, and purposeful humane growth in the society, particularly in view of threats from climate change.

As change is the only consistent phenomenon in this universe, one should remain prepared to challenge her/his past achievements, and transform accordingly. While doing so creating a fearless, nature-centric enabling environment for "all" is equally important. The Earth, water, light, air, fruit, flower, and so on are not simple physical phenomena for use, but these are the notes to complete the symphony of joy, peace, and fulfillment. The perfect comprehension of the integrated balance between the human and nature can be best expressed, in Tagore's words—in freedom of mind and fulfillment of responsibilities. Hence, our behavior must be in consonance to this line of philosophy.

The idea to write this book was germinated while conducting a series of interactive workshops and training programs for the Malé-based SAARC Coastal Zone Management Centre during the early years of this decade. The outcome confirmed the perceived threat of this region from the climate change impacts. Although climate change is a global issue, localized solutions have become increasingly necessary to address political, economic, and cultural factors in a region like South Asia. In this regard, inadequate infrastructure, staggering illiteracy, extreme poverty, and lack of hygiene in this region could make things worse. Additionally, lack of trust among the nations and increasing menace of fanatic-fundamentalism would make any global action to repair the situation in South Asia difficult and at times, counterproductive.

The growing aspiration of more than one billion youth of South Asia with no commensurate employment opportunity could become a problem not only for this region alone but for the entire world. While identifying success, gaps, and shortcomings in existing policies and regional laws relating to climate change, this book evaluates the sustainability of current practices.

Probably the entire world has gone far away from the truth—that superiority of human being does not rest in power of possession but in strength of unison and empathy. The greed that largely drove the market economy during the last two centuries needs to be shackled in. The Buddha-Tagore-Gandhi (BTG) doctrine and the BTG Wheel of alternate governance model are proposed here to circumvent this enormous predicament, even partially.

This book has six chapters. We initiate by discussing the philosophy of climate change in the first chapter. Climate essentially rests on the global energy balance and emission scenario. The balance is expressed as the difference between the total energy received from the Sun, against the sum of energy reflected by the clouds, and that absorbed by the Earth and the atmosphere. The imbalance of the above since the industrial revolution in the 18th century witnessed an unmatched increase in the concentration of greenhouse gases (GHG) in the atmosphere and the temperature of the planet Earth. Although emitting lowest per capita CO_2 (a member of GHG), South Asia is going to receive the maximum brunt of climate change, in terms of agricultural yield, health, rainfall, drought, and severe other climatic events. Moreover, South Asia suffers from extreme poverty, inadequate health, stunted infrastructural facilities, and economic, social, and educational backwardness. Additionally, this region is the hotbed of illiteracy, corruption, and religious extremism. Yet, South Asia can hardly afford to lose this battle against climate change.

The second chapter documents the marked surge in emission of greenhouse gases (GHG) in the atmosphere over the last 150 years. GHG creates a suffocating greenhouse effect and drives the temperatures up by several degrees. This chapter lists the emission and concentrations of all major GHG and discusses their impact on atmosphere, ocean, land, and humanity. The chapter ends with a rapid appraisal of the entire climate change paradigm.

Carbon politics at the global and regional levels are delved in the third chapter. Climate change may revolutionize our cognition, attitude, and behavior. While threats from change in climate should enhance international cooperation and collaboration to avert or reduce impact, the entire stage is vitiated by the self-conceived pride of various countries, bordering self-denial. A plethora of conflicts of interest has divided the world into two schools and three worlds. Paris Agreement in 2015 (and follow-ups) witnessed some appreciable achievement to reduce these polarities after almost two decades of fruitless negotiations.

Some consider climate change as the outcome of the greatest market malfunction, representing uneven discriminatory growth the world had witnessed during the last two centuries. Such noninclusive asymmetric development makes it necessary for a region like South Asia to formulate public schemes to reduce the impact of climate change, quickly and squarely. While all the eight South Asian countries have chalked out their adaptation and mitigation strategies, an element of disconnection between policies and people is very distinct. Rational education, mutual trust, cooperation, and integration of markets and economies could bring about "achhe din" (good days).

Climate change need not always be a threat. It can be viewed as an opportunity. The concept of transforming climate change threats to opportunities, including job creation, has recently started emerging in South Asia. The fifth chapter examines how the threat of climate change on the humanity could be transformed into opportunities. Such a transformation, if successful, could bring sustained and inclusive growth not only in economy, but also in the society. The present trend in innovations and morality while undertaking climate engineering came under special discussion, confirming that prevention (mitigation) is better than living with it (adaptation), and cure (climate engineering).

The last chapter prescribes an alternate governance model (Buddha-Tagore-Gandhi doctrine) for South Asia to optimally mitigate the impact of global warming and climate change (GWCC). While doing so, emphasis is given on the transformation in mindset, and a change in values of the people, government, and entrepreneurs. Anomalous increase in population and swelling of selfish-greed needs to be controlled quickly. Reforms in agricultural and water management sectors and in food habit are also discussed. The BTG doctrine involves repairing the intrinsic relationship between human and the nature, encouraging in the process ethical innovation, gender equality, women empowerment, and sustainable development. In consonance to the underlying philosophy of "Think Global, Act Local," the BTG doctrine can be seen as a game-changer for the other areas also.

The work of this book was carried out entirely in the campus of the Goa-based CSIR-National Institute of Oceanography, one of the finest laboratories in this part of the world. The logistic and library support have been wonderful and excellent. We acknowledge these helps from bottom of our hearts. Much of the credit for this immensely rewarding intellectual experience must also go to Professor Anil K Ghosh, and a young engineer cum business manager—Ayan. Interacting with them has developed our analytical skills more than what we could have imagined and enhanced organization of thoughts and clarity in expression.

We place on record our special thanks to the Elsevier team—Candice Janco (*Publisher*), Laura S Kelleher (*Acquisition Editor*), Tasha Frank (*Editorial Project*

Manager), Moosa Omer Mukhtar (*Production Project Manager)*, and Aswathi Aravindakshan for their constant help throughout this venture. This book would not have seen the light of day without their patient and cool handling of several awkward situations. Matthew Limbert designed the book.

Dr. S Mandal (formerly with CSIR-NIO) provided part financial support from his project toward meeting expenses for acquisition and processing of critical data from various sources. Several rounds of discussion (*bordering heated arguments*) with Chanchal Dass (an excellent entrepreneur-engineer) guided our thoughts throughout. Baishali Gupta and Prof. Dolly Mathew corrected our approach while commenting on the first draft. Much credit goes to these four peer-members. Their feedbacks have been invaluable.

We thank a host of our colleagues and students for helping us in organizing the book matter at several stages. They include Rohit Kerkar, Nilovna Chatterjee, Sachit Kuttikar, Aaheli Bhattacharya, Davis Thomas, Gabriella D'Cruz, Akeek Maitra, Ankita Dutta, Mohammed Irfan, and Poornima Dhawaskar. Dr Karisiddaiah specially thanks his wife, Shailaja, for her constant encouragement and patience throughout this book-writing period. We also thank our other family members and friends in Goa, Kolkata, and Bangalore for their tolerance and indulgence.

May 14, 2018

Ranadhir Mukhopadhyay
CSIR-National Institute of Oceanography

S M Karisiddaiah
Centre for Contemporary Research
(formerly with National Institute of Oceanography)

Julie Mukhopadhyay
Ganga Zuari Academy

CHAPTER 1

Introduction

1.1 PHILOSOPHY OF CLIMATE CHANGE

The physics and chemistry of the Earth's Troposphere (ground to 11 km altitude) largely determines the climate (Lockwood, 1979). Any alteration in such physicochemical environment causes climate change, a term first coined in 1966 by the World Meteorological Organization to encompass all forms of climatic variability on timescales longer than ten years. The climate-equilibrium of the Earth depends on the difference between the total energy received from the Sun and part of the energy that is returned to the outer space. Of the total solar radiation of 341.3 W received on the Earth annually in every square meter (W/m^2), about 102 W/m^2 is reflected back by the clouds, aerosols, atmospheric gases, (and some even by the Earth), while the rest 239 W/m^2 is absorbed and shared between the atmosphere (78 W/m^2) and the Earth's surface (161 W/m^2, Fig. 1.1; Houghton et al., 2001; NASA, 2014; Kiehl and Trenberth, 1997).

The GHG in the atmosphere are a family of gases that absorb and emit radiation within the thermal infrared range. When absorbed in the lower atmosphere, the GHG create a suffocating umbrella effect (the greenhouse effect) by not allowing a considerable portion of radiation to return to the atmosphere from the Earth. In the process, GHG warms the globe and sets in climate change (Fig 1.1, Walsh et al., 2014a,b).

Although the combined GHG constitute only about 0.03% of the atmospheric volume (the other being nitrogen: 78.09%, oxygen: 20.95%, and argon: 0.93%), they have a profound impact on the Earth's climate. The GHG in the atmosphere are contributed largely by human activities. Its major components are (a) carbon dioxide (CO_2), which is emitted through combustion of fossil fuel, deforestation, and desertification, (b) methane (CH_4), which is discharged through rice cultivation, cattle rearing, biomass burning, coal mining, ventilation of natural gases, landfill, and wood fuel use, (c) nitrous oxide (N_2O), which is released through agriculture, fossil fuel combustion, and use of catalytic converters in cars, and (d) chlorofluorocarbon (CFC), which is produced by airconditioners, freezers, solvents, and insulators.

Climate Change. http://dx.doi.org/10.1016/B978-0-12-812164-1.00001-3

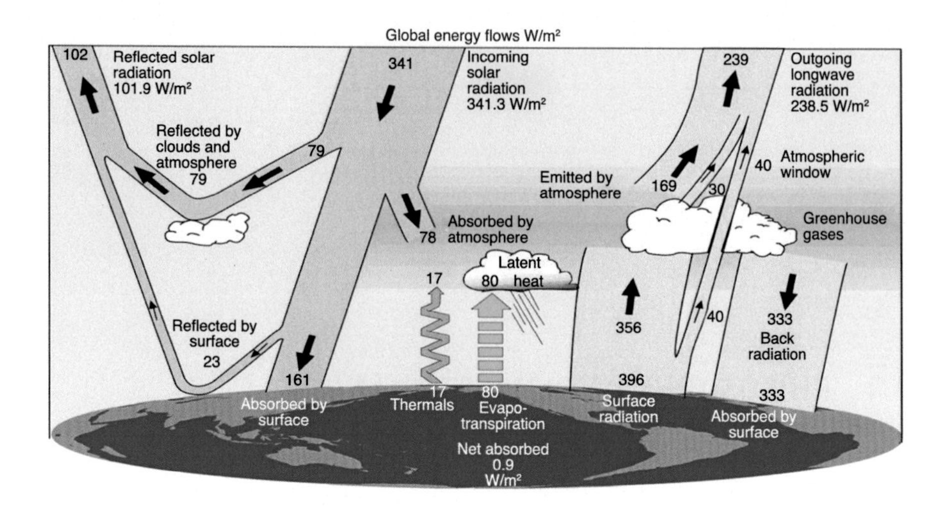

FIGURE 1.1 Earth's annual global mean energy budget (tentative).
(Modified after NASA 2014. Warmest year in modern record, Global Climate Change, National Aeronautics and Space Administration, USA; Kiehl, J.T., Trenberth, K.E., 1997. Earth's annual global mean energy budget. Bull. Am. Meteorol. Soc. 78, 197–208; Houghton, J.T., Ding, Y., Griggs, D.J., Noguer, M., van der Linden, P.J., Dai, X., et al., 2001. Climate Change 2001: The Scientific Basis. IPCC, Cambridge University Press, Cambridge, pp. 39; IPCC. Climate Change 2014: Impacts, Adaptation, and Vulnerability. Part A: Global and Sectoral Aspects. In: Field, C.B., Barros, V.R., Dokken, D.J., Mach, K.J., Mastrandrea, M.D., Bilir, T.E., Chatterjee, M., Ebi, K.L., Estrada, Y.O., Genova, R.C., Girma, B., Kissel, E.S., Levy, A.N., MacCracken, S., Mastrandrea, P.R., White, L.L. (Eds.), Contribution of Working Group II to the Fifth Assessment Report of the Intergovernmental Panel on Climate Change. Cambridge University Press, Cambridge, UK, New York, NY, USA, pp. 1132).

The global atmospheric concentrations of CO_2, CH_4, N_2O, and other GHG have risen significantly over the last few hundred years. For example, CO_2 concentration rose from 280 ppm in the year 1700 to 401 ppm in 2015, CH_4 from near zero to 1800 ppb now, and N_2O from 280 ppb to 328 ppb during the same period (US-EPA, 2016). On May 9, 2013, CO_2 levels in the atmosphere touched 400 parts per million (ppm; or 0.04% of the atmosphere; NOAA-SIO, 2013). The CO_2 emission was found to be rising at a little over half a ppm annually since the 1960s.

The emission of CO_2 has the potential to increase the temperature and warm the globe. Consequently, the average temperature of the Earth's surface (average 15°C) has risen by about 0.8°C in the last 100 years. Of this, about 0.6°C increase has been recorded since the last three decades (Table 1.1). Satellite data further shows that the level of sea has elevated by about 3 mm/year in recent decades, caused largely by the thermal expansion of seawater. This happens as seawater warms up and loosens the packing of water molecules, causing an increase in its volume. In addition, glaciers of Antarctica have been melting since 1979 at an annual rate of 4% (IPCC, 2007, 2014).

Table 1.1 Variations in GST Over Land and Sea During the 20th Century

Year	January	February	March	April	May	June	July	August	September	October	November	December	Average
Land surface mean temperature													
1901–2000	2.8	3.2	5.0	8.1	11.1	13.3	14.3	13.8	12.0	9.3	5.9	3.7	8.5
Sea surface mean temperature													
1901–2000	15.8	15.9	15.9	16.0	16.3	16.4	16.4	16.4	16.2	15.9	15.8	15.7	16.1
Combined mean surface temperature													
1901–2000	12.0	12.1	12.7	13.7	14.8	15.5	15.8	15.6	15.0	14.0	12.9	12.2	13.9

Temperatures are given in centigrade.
Source: NCDC-NCEI-NOAA (2016).

In fact, the climatic situation changed drastically by the middle of the last century, when a series of calamities engulfed the world. For example, the ozone layer was depleted due to the accumulation of Freon caused by aviation transport, the disease of cancer attributable to the increased use of chemicals, extinction of several avian species owing to pesticides used in agriculture, and loss of forests because of acid rain (Idso and Singer, 2010).

Change in climate is likely to threaten all forms of life on the Earth. The impact will be felt more on people's health, agriculture, forestry, water resources, coastline modification, sea-level changes, human migration, psychological behavior, industry, energy, and nation's economy (Bryson, 1997; Lovejoy and Schertzer, 2013; Pielke, 1998). The COP-21 (Paris Agreement) rightly calls for reduction in global emission of GHG to keep the warming within the threshold of 2°C by adopting innovative and responsible low-emission renewable technology for generation of energy. Moreover, global warming and related change in climate have even the potential to lead to a violent social upheaval (implosion) in the future, if not addressed promptly and appropriately.

1.2 PAST CLIMATE CHANGES

It is wrong to assume that climate change is a new phenomenon. In fact, there have been at least five significant falls in temperature in the Earth's history. Actually, a large portion of the Earth was covered by thick ice sheets during the Paleo-Proterozoic time (Huronian, 2400–2100 Ma, or million years before present), Neo-Proterozoic (Cryogenian, 850–635 Ma), Silurian-Ordovician (Andean-Saharan, 460–430 Ma), Permian-Carboniferous (Karoo, 360–260 Ma), and Quaternary (2.6 Ma to today; Fig. 1.2, Table 1.2) periods. Furthermore, approximately a dozen epochs of glacial spreading occurred during the past one million year—the largest of which peaked 650,000 years ago and lasted for about 50,000 years. The most recent glacial period culminated some 18,000 years ago before giving way to the interglacial Holocene epoch

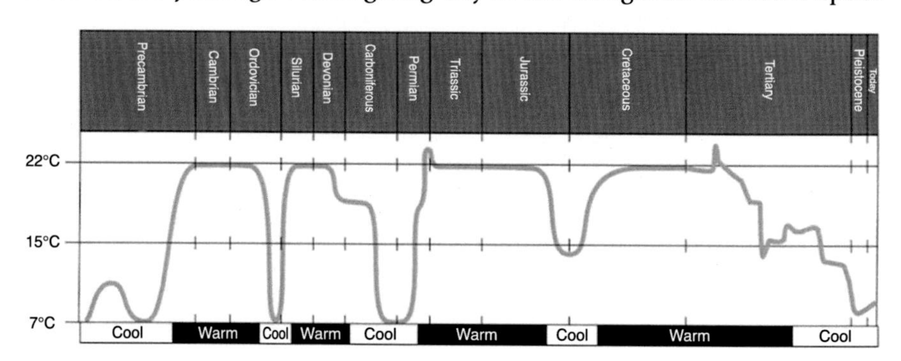

FIGURE 1.2 Variations in global temperature since the Precambrian times.
(Modified after Scotese (2015)).

Table 1.2 Ice Ages in Geological History

Name	Period (Ma)	Period	Era
Quaternary	2.58–present	Neogene (and Quaternary)	Cenozoic
Karro	360–260	Carboniferous and Permian	Paleozoic
Andean-Saharan	450–420	Ordovician-Silurian	Paleozoic
Cryogenian (Sturtian/Varanganian)	720–635	Cryogenian	Neoproterozoic
Huronian	2450–2100	Siderian and Rhyacin	Paleoproterozoic
Pongola	2900–2780		Mesoarchean and Neoarchean

Source: Modified after Walker et al. (2009) and Kopp et al. (2005).

Table 1.3 Quaternary Glacial Cycles That Covered Large Portions of the World With Ice

Ice Stage	Period (ka)	Marine Isotope Stage (MIS)	Epoch
Interglacial	Present–12	1	Holocene
Glacial	12–110	2–4 and 5a–d	
Interglacial	115–130	5e	
Glacial	130–200	6	
Interglacial	374–424	11	Pleistocene
Glacial	424–478	12	
Interglacial	478, 533–563	13–15	
Glacial	621–676	16	

ka, *Kilo years.*
Modified after *Gibbert and van Kilfschoten (2004) and NEEM (2013)*
Source: Encyclopedia 2017.

11,700 years ago (Table 1.3). Interestingly, the Ice Age occurred much before the human advent on this Earth, and naturally much before the GHG emissions contributed by the human beings.

At the height of the Quaternary glaciation, the ice grew to 3–4 km thick, as sheets spread across Canada, Scandinavia, Russia, and South America. Correspondingly, sea levels lowered to more than 120 m, while global temperatures dropped to around −12°C on an average. The ice cover caused substantial change to the landscape through erosion and deposition. Such changes may have forced plant migration, drop in sea levels enabling rivers to excavate deeper valleys, and produced enormous inland lakes, thereby exposing submerged land bridges. During deglaciations, the melted glacier water may also have created new lakes.

Crutzen and Ramanathan (2000) traced the history of climate and atmospheric sciences. The fundamental theory of Greek philosopher Aristotle on climate had remained unchallenged for nearly 2000 years until the 17th century, when Joseph Black first identified CO_2 in the air in 1750 and Henry Cavendish in 1781 measured nitrogen and oxygen present in air. The greenhouse effect, which is a natural process, was first recognized by the French mathematician and physicist Joseph Fourier (Fourier, 1824) and confirmed by British scientist John Tyndall in a series of experiments, starting in 1859 (Tyndall, 1861). Tyndall suggested that water vapor, CO_2, and other gases are keeping the Earth warm. The Swiss geologist Louis Agassiz and Serbian mathematician Milutin Milankovitch first addressed the causes behind ice ages (glaciations/deglaciation). While Agassiz (1841) identified rock striations and sediment piles as evidence of glacier activity, Milankovitch (1941) predicted the cyclical variations in Earth's eccentricity, precession, and obliquity that are likely to determine the amount of solar energy (solar insulation) received by the Earth. Shifting of tectonic plates due to Earth's internal heat transfer also contributed to the thermometry of the globe. In fact, as late as 1814, a Frost Fair was to be held over the river Thames, in London, which used to freeze every winter.

For the first time in 1959, satellite images of cloud cover were made available to help estimate the global radiation budget. During 1970–74, several studies on CFCs damaging ozone layer were published. During the nineties, the cooling effect of atmospheric aerosols and their importance in offsetting the greenhouse effect was discovered.

1.3 GLOBAL EMISSION SCENARIO

A positive relation appears to exist between carbon emission and strategic power index (SPI) of a nation, in a way similar to the gross domestic product (GDP) of a country reflecting its economic strength. Thus a decrease in CO_2 emission would result in a commensurate decline in prosperity, because prosperity needs energy, and GDP closely correlates to such prosperity (Langmuir and Broecker, 2012). For example, a study on economic growth in Nigeria (a unique country from climate change impact angle) from 1970 to 2013 suggests that economic growth, trade openness, and capital investment positively impact carbon emission. Yet, a reduction in GDP in an attempt to curb carbon emission can harm the country's economic progress. Therefore, the nation must look for alternate ways to promote green growth in the country (Mesagan, 2015).

It is observed that the country that has enhanced energy production records higher GDP, consequently achieving two-to-three-fold prosperity, becoming more powerful economically, and strategically influencing regional and global affairs. Now, which strategically powerful country on this Earth would like to lose this power and status? And because North America and Europe emitted more CO_2 between 1800 and 2000 to ensure a better living standard for their citizens, how could China and India be stopped now if they want to provide a higher living standard for the betterment of their people? Hence, a fundamental conflict exists between increasing prosperity and reduction in CO_2. The answer to achieve the twin goals of inclusive growth—the Millennium Development Goals (MDG) and the Sustainable Development Goals (SDG)—lies in focusing on increased investment in human capital (education and health), and steps to make development efforts environmentally responsible (World Bank, 2015).

The recent data suggests that top ten CO_2 emitting countries in the world produce around 70% of global GHG emissions (WRI, 2015). The uncertainty in CO_2 emissions is generally low (below 10%). However, the same cannot be said for CH_4 and N_2O emissions in view of lack of data (Janssens-Maenhout et al., 2017). In fact, USA, Canada, and Australia top the world with highest per capita GDP, highest per capita energy use, and highest per capita CO_2 emissions. The UK, Japan, New Zealand, Western Europe, and surprisingly, the Middle East hog the intermediate portion of this per capita triangle of GDP, energy use, and CO_2 production. Quiet expectedly, India, China, the rest of Asia, Africa, and Latin America are found at the bottom of this "pyramid of affluence." Despite comprising a meager 5% of the world population, the USA has emitted the most CO_2 over the last two centuries (IEA, 2016). Yet, per-capita emission of CO_2 by China is still less than half of that of the USA, and that of India remains nine times less than the USA.

The level of GHG emissions divided by GDP is a commonly used measure of emissions intensity for a country. This data remains advantageous when looking at the decarbonization of the national economy or energy system. Yang et al. (2017) constructed a comprehensive reduction index to allocate the carbon emission quotas among different industrial sectors fairly and effectively. The emission intensity for the top 10 emitters of GHG vis-à-vis GDP is estimated. For the energy sector, the world average is 410 tons of GHG emissions (CO_2) per million-dollar GDP, but intensities vary across countries.

Seven of the top 10 emitters admittedly have an emission intensity that is below average. However, Russia, China, and Canada are above the world average. These differences not only result from varying emission levels and size of economy, but are also dependent on type of carbon-intensity sectors. However, if emission from deforestation and land-use changes is taken into account, Indonesia may as well become the most intensive emitter. For a real-time comparison, the CO_2 emission during the year 2014 in 44 countries and CO_2 presence between 1970 and 2013 are estimated.

The simulated global temperature (Fig. 1.3) demonstrates that the natural forces (blue line) and human influences (pink line) together have been responsible for the increase in global temperature, substantially by at least 1 °C during the last century. The black line is the global average temperature as

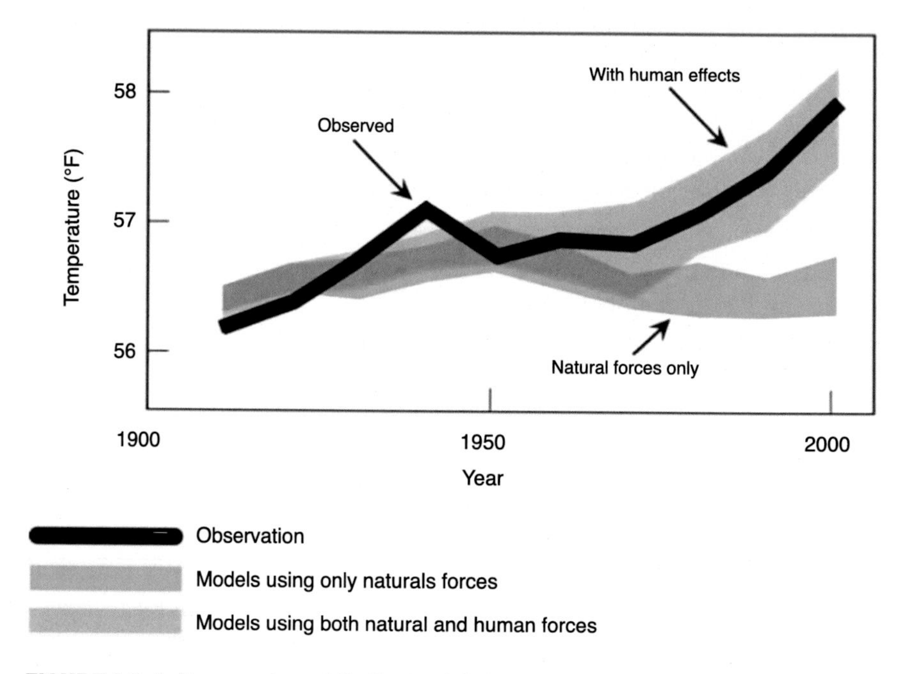

FIGURE 1.3 Anthropogenic contribution to global temperature increase.
(Modified after IPCC Climate Change 2007: Synthesis Report. Contribution of Working Groups I, II and III to the Fourth Assessment Report of the Intergovernmental Panel on Climate Change (Core Writing Team, Pachauri, R.K., Reisinger, A. (Eds.)). IPCC, Geneva, Switzerland).

actually observed. The close match between the black line and the pink line indicates that observed warming over the last half-century cannot be explained by natural factors alone, and may have been augmented by the human factors (NOAA, 2016). The threats of South Asia, sheltering about 23% of world population in a 3.4% of world land area, will naturally be huge as enunciated by an econometric model explaining the causal relationship between carbon dioxide emissions and population (Sikdar and Mukhopadhyay, 2016).

Again, the increase in atmospheric abundance of GHG and a concomitant decrease in the concentration of ozone in the stratosphere have resulted in increased UV-B radiation on the Earth. Ozone is of two types—the one found in stratosphere (about 9–48 km above the Earth's surface) is good ozone preventing harmful UV radiation from the Sun to reach the Earth. However, ozone accumulated closer to the earth (at troposphere) is harmful, and creates pollution and smog (US-EPA, 2016).

Globally, it is found that the annual CO_2 emissions from the burning of fossil fuels have reached an alarming 32 billion tons. This was enhanced further by an additional input of about 4 billion tons from deforestation. Because the Earth can absorb only 17–18 billion tons every year, the atmospheric CO_2 levels have increased three times faster than the average during the last decade. This is a several thousand times more than the natural rate at which CO_2 gets in and out of the atmosphere, as part of the carbon cycle (IPCC, 2014; NOAA, 2016).

As the researchers struggle to project how fast, how high, and how far the oceans will rise, the IPCC (2007) projected that sea levels would rise between 28 and 97 cm by the last decade of this century. This upper limit seemed far too low to other scientists, given the pace of melting of ice in Greenland (Meehl et al., 2013). The semiempirical model of Church et al. (2011) investigated sea level rise since the 1880s in correspondence with air temperatures, and predicted a 140 cm rise of sea level by 2100, more than the IPCC prediction. Any increase in CO_2 level (consequently temperature) would cause Arctic ice to melt first. Another estimate, however, concluded that half of the apparent sea level rise was attributable not to rising waters but to sinking land (Idso and Singer, 2010).

Ocean, incidentally, has been the best "sink" for atmospheric $CO_{2,}$ and absorbs about 30%–40% of the same (Hardy, 2003). As a consequence, the oceans are turning more acidic, thereby imposing a threat to the entire marine life. Hence, CO_2 acts as climate regulator and deserves a careful handling as the Earth moves in and out of the Ice Age (glacial age). One must be cautious in fiddling with this regulator (Adve, 2013; Archer et al., 2009).

The Earth's current trend of GHG emissions is as per the Paris Agreement, but this may fall short of stabilizing the climate system below 2°C above preindustrial levels. It is claimed that to bring back Mother Earth to its pristine state before the advent of civilization, it is important to reduce the concentration of CO_2 in the atmosphere to 350 ppm or less (Hansen et al., 2008). Not everything about GHG is bad. It is also presumed that without the natural greenhouse effect, the average surface temperature of the Earth would have been about 15.55°C cooler. The concentration of GHG and its impact particularly on South Asia is assessed in the next section.

1.4 WHY SOUTH ASIA

Beyond all these intricacies of data and hypothesis, it seems that climate is changing, slowly but certainly. However, the rate and cause of such change sparks many debates. Moreover, impact of climate change on human life and society is yet to be gauged accurately. The same is more important for a region like South Asia, which holds about 1.749 billion people (~23% of world population) in an area of about 5.1 million km^2 (~11.51% of the Asian and 3.4% of the world's land surface area). Extreme poverty, inadequate health facility, diminutive infrastructure, and economic, social, and educational backwardness exacerbate the condition of South Asia. In fact, approximately 40% of the world's poor live in rural regions of South Asia. Therefore, a significant gain in global poverty reduction target can be achieved if focus is made on South Asia (Thapa, 2004).

Considering these factors, it would be reasonable to assess the impact of global warming and climate change on South Asia. It is also deemed necessary to consider the impact of cataclysmic events like heat waves, droughts, floods, hurricanes, and sea level rise on the health, education, agriculture, water resource, hunger, and infrastructure issues in South Asia. Further, it would be interesting to study the way governments in this region would deal with such existential situations, and the urgency to develop an appropriate public policy based on scientific analysis. An assessment of the types of innovative adaptive measures that could be taken up in conformity to the ethics, culture, and customs of this region, and inculcating a sustainable, need-based (not greed-based) grass-root approach, should be desirable.

This book especially focuses on South Asia, comprising eight countries—Afghanistan, Bangladesh, Bhutan, India, Maldives, Nepal, Pakistan, and Sri Lanka (Fig. 1.4)—which also form the South Asian Association for Regional Cooperation (SAARC). South Asia has a distinct geographical identity, and the region is indigenous to a variety of geographical features, such as mountains, glaciers, rain forests, valleys, deserts, rivers, lowlands, plateau, and alluvial fans. It is surrounded by three water-bodies—the Bay of Bengal, the Indian Ocean, and the Arabian Sea.

South Asia is being crowned by the mighty Himalayas to the north, which blocks the north-Asian bitter cold winds from blowing to the south, thereby keeping the temperatures considerably moderate in the plains down south. The monsoon that brings rain to South Asia keeps the region humid during summer and dry during winter. The climate favors the cultivation of jute, rice, tea, cereals, and vegetables. During the summer monsoon, wind blows from southwest to most parts of the region, accounting for 70%–90% of the annual precipitation. The Indian Ocean provides the principal connection between the Pacific, Atlantic, and Antarctic Ocean basins and between the upper and lower layers of the global ocean circulation. Therefore, it is appropriate to say that changes in the waters of the South Asian nations (i.e., northern Indian Ocean) will have a global impact.

South Asia can be largely divided into four broad climatic zones—(a) the northern mountains having a dry subtropical Alpine climate, (b) the river lowlands with a tropical semi-arid climate (hot summer and cool winter), (c) the plateau far south of India and southwest Sri Lanka with an equatorial climate, and (d) the hot subtropical climate in northwest India, deserts in Rajasthan, and northwest of Pakistan. Even a minor variation in climate can have significant impacts on the society, economy, industry, and agriculture of these countries; and on lifestyle, food habit, and health of individuals. The variation in climate in South Asia is influenced not only by the altitude, but also by factors

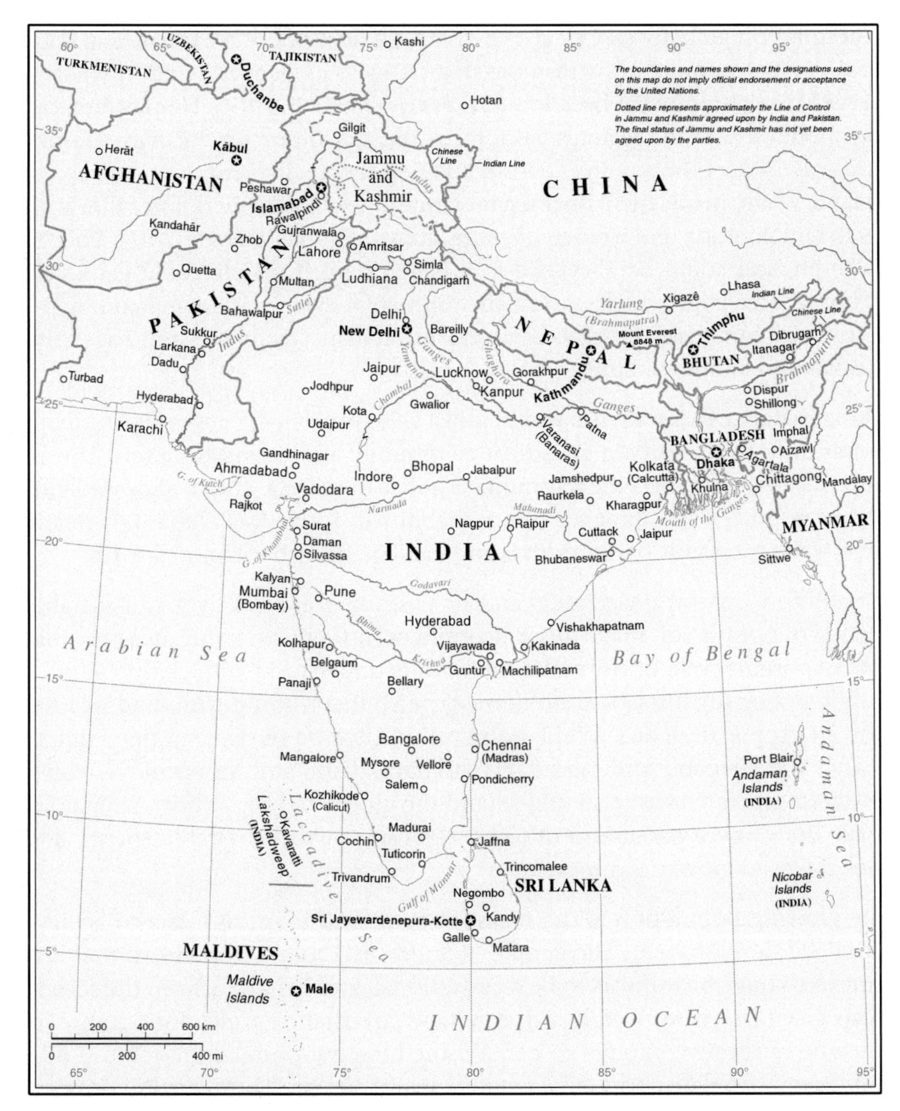

FIGURE 1.4 South Asian countries.
By UN (UN.org) via Wikimedia Commons.

such as proximity to the seacoast and the seasonal impact of the monsoon (Burtman and Molnar, 1993). Southern parts are mostly hot in summers and receive rain during monsoon periods. The northern areas of the Indo-Gangetic plain remain hot in summer, but cool in winter. The mountainous north is cooler and receives snowfall at higher altitudes of the Himalayas.

Although South Asia is not emitting much per-capita carbon, this poverty-ridden region is threatened to be significantly impacted following changes in carbon concentration globally (IPCC, 2014). This is because South Asia shelters a huge pool of rural population with more than 70% of its people living in villages. Growth in agriculture has been low (2% per capita, ADB, 2014). Despite employing about 60% of the region's labor force, the agriculture productivity contributes only 22% of the region's GDP. In addition, unequal distribution and access to land, where many are neither tenant farmers nor farm owners, force the farmers to remain poor, and women insecure and vulnerable (Sterrett, 2011). Poverty in South Asia could be alleviated to a large extent if people could get access to improved energy services, arrest environmental deterioration and curb widespread religion-based misplaced violent extremism (Bierbaum and Fay, 2010; Casillas and Kammen, 2010).

Climate change in South Asia would inflict wide variety of changes among mangroves, fish, coral, and sea grasses, at community and ecosystem levels. This in turn will adversely affect the way humans utilize the coastal and riverine wetlands and its resources. The infrastructure availability in South Asia (Table 1.4) clearly depicts the economic backwardness and inadequate technological growth.

In addition, several unscientific archaic customs followed by a considerable section of population make this subcontinent further vulnerable. It is time that this important cycle of cultural dimension is understood with clarity before a public policy for infrastructure development, threat mitigation, and adaptation strategy is designed. While doing so, it must be recognized that climate change is irreversible and caused by both natural and anthropogenic activities. Furthermore, one must also understand the global carbon politics, as nations make use of new generation of technological growth, renewable energy, and innovative adaptive measures.

The growing population is the major impediment for South Asia to achieve sustained development. The people of South Asia may comprehend that it is time to change the mindset to be more rational, and help transform the society from greed-based to need-based. Any new governance policy for South Asia must be innovative enough to contain the bloating population, remove economic and social disparity, and augur inclusive growth. The pivotal to this new policy may be cooperation and collaboration of all the South Asian countries, at least for the next two decades. The countries should wholeheartedly take part in this campaign with equal emphasis on ethics, values, and egalitarianism to enable sustainable, balanced, and purposeful growth in the region.

In short, this book will provide an assessment of climate change issues through the socioeconomic lens of one of the world's poorest and most populous regions. Although climate change is a global issue, localized solutions have become increasingly necessary to address political, economic, and cultural factors. The sustainability of current practices will be evaluated, and a responsible innovative alternate model of governance will be discussed.

Table 1.4 Infrastructure Availability in South Asia

	Access to Telephone, 2013	Access to Electricity, 2012	Access to Improved Water, 2012	Access to Improved Sanitation, 2012	Internet Users, 2013	Rail Density, 2012	Road Density	Paved Roads, 2012
	a	**b**	**c**	**d**	**e**	**f**	**g**	**h**
Afghanistan	74	–	55	32	6	–	35	36
Bangladesh	76	60	87	61	10	22	1,838	10
Bhutan	85	–	100	50	34	–	219	34
India	75	75	94	40	18	22	1,578	54
Maldives	192	–	99	98	49	–	293	100
Nepal	85	76	92	46	15	–	139	54
Pakistan	76	69	91	64	14	10	341	73
Sri Lanka	120	85	96	95	26	23	1,819	15
South Asia	76	73	92	45	17	19	1,123	52
E & NE Asia	118	98	96	80	53	8	400	64
SE Asia	132	77	90	72	29	5	276	55
World	111	78	91	67	40	9	275	57

a, e: per hundred people; b, c, d, h: percent of population; f, g = km/1,000 km^2; E: East; NE: Northeast; SE: Southeast.
Source: From Achieving the SDGs in South Asia: Key Policy Priorities and Implementation Challenges, by Michael Williamson, © 2017 United Nations. Reprinted with the permission of the United Nations.

CHAPTER 2

Scientific Assessment

2.1 CAUSES OF CLIMATE CHANGE

As indicated earlier, climate change is caused both by natural events and anthropogenic activities. We discuss these below.

2.1.1 Natural Causes

A number of natural factors influence the ocean and atmosphere periodically and greatly impact the climate. Few among these are drifts of continents, volcanic eruptions, ocean currents, variations in the Earth's orbital characteristics and solar output (Zacharias, 2008). In fact, the transport of heat and GHG from atmosphere to land, land to ocean and from ocean back to the atmosphere affect the evaporation and precipitation patterns, and alter the productivity of planktons, invertebrates, and fish. Such changes can not only cause drought and floods, but also could affect forestry, agriculture, and economy (Hardy, 2003). Some of the natural causes for global warming are briefly touched upon here.

2.1.1.1 Sunspots and Solar Radiation

The solar radiation deriving from thermonuclear reaction on the Sun reaches Earth in three wide bands of electromagnetic spectra—short wavelength (X-rays, ultraviolet), medium wavelength (visible light), and long wavelength (infrared). The GHG allows the sunlight to reach the Earth but traps the outgoing infrared long wavelength radiation. Held back radiations alter the net heat balance of the Earth. The atmospheric and oceanic currents carry this extra heat from the tropics to the polar regions and remain responsible for increased melting of ice.

Solar magnetic fields produce sunspots, whose number increases and decreases with a 10.7-year periodicity (Milankovitch Cycle). This puzzling regularity in the Sun's activity is known as solar cycle. During the deep solar minimum, characterized by a lower frequency of solar storms, the Sun's magnetic field weakens. This allows cosmic rays to penetrate the solar system in record numbers, making the outer space a more dangerous place. At the same time, the

Climate Change. http://dx.doi.org/10.1016/B978-0-12-812164-1.00002-5

decrease in ultraviolet radiation makes the Earth's upper atmosphere to cool and collapse (Nandy et al., 2011).

2.1.1.2 Sea Surface Temperature

The natural ocean conveyor belt (NOCB) passes through the world's oceans and helps regulate the planet's climate. For example, NOCB carries warm waters from the tropics to the North Atlantic and from the east Pacific Ocean to the Indian Ocean, making places like Iceland and Western Europe warm enough to be comfortable and generating monsoon in South Asia, respectively.

In fact, at times, the sea surface temperature (SST) would become warmer or cooler. Such deviations may occur at irregular intervals of roughly 3–6 years along the coasts of Ecuador and northern Peru in the eastern Pacific Ocean. This results in initiating El Nino (warm) and La Nina (cool) climate cycles, respectively. The impacts of El Nino and La Nina are hence opposite to each other, and are found to influence weather patterns across the globe. The El Nino (also known as El Nino Southern Oscillation, ENSO or Pacific Warm Episode, PWE) involves change in ocean currents and associated heat transport from high atmospheric pressure center in the eastern Pacific to a low-pressure center in Indonesia and northern Australia.

Normally, this high-pressure difference between the eastern and western parts of the Pacific Ocean drives trade winds from east to west along the equator. This westerly blowing wind forces warm seawater to pile up in the west (rising sea level by about 40 cm in the Indo-Pacific region) and depresses the thermocline to about 200 m deep. A thermocline is the transition layer between warmer mixed water at the ocean's surface and cooler deep water below. In the thermocline, temperature decreases rapidly from the mixed upper layer of the ocean (called the epipelagic zone) to much colder deep water (mesopelagic zone). The warm water causes increased evaporation and high rainfall in the west (Indo-Pacific) and dry air (drought?) over South America (Rodbell et al., 1999).

In contrast, the La Nina (opposite of El Nino) makes SST in tropical Pacific drop below normal (Hardy, 2003). Consequently, the pressure difference between the Indian Ocean and the Pacific Ocean becomes low, and a weak westerly trade wind fails to transport adequate heat toward the west. The warm water of the west now flows back toward the east, thus raising the sea level and causing heavy rain in east the Pacific region and drought like situation in the west (Indo-Pacific region). Any modification to ENSO and La Nina therefore, could disturb the intensity of monsoon in the Indian Ocean, affecting in the process agriculture and economy of the entire stretch of Afro-Asian countries

2.1.1.3 Volcanic Eruptions

Volcanic eruptions and global warming are inextricably related since the formation of the Earth. This is because volcanoes have been spewing enormous amount of ash, water vapor, and GHG into the atmosphere. These expulsions of gases along with hot lava that burns the vegetation around can noticeably change the climate, often in devastating ways in the geological history.

For example the major disappearance of organisms from the Earth occurred five times during the last 500 Ma (million years, Alvarez, 2003; Coffin et al., 2006). These extinction phenomena corresponed to large volcanism at the boundaries between geological periods—Ordovician and Silurian (at 443.7 Ma), Devonian and Carboniferous (359.2 Ma), Permian and Triassic (251 Ma), Triassic and Jurassic (199.6 Ma), and Cretaceous and Tertiary (at ~65 Ma). These volcanisms formed large igneous provinces in the world, such as the Deccan Traps in western India at 65 Ma. Considerable amount of erupted hot lava caused abrupt increase in GHG, setting in prolonged darkness and extreme cooling, episodic oceanic anoxic events, and repeated marine transgression, leading to a major biotic catastrophe. Working together, the aforementioned events could bring about sustained sharp change, in climate making it inhospitable and hostile for large soft-bony reptiles and other organisms to survive (Bond and Wignall, 2014; Chatterjee et al., 2013; Percival et al., 2015; Self et al., 2014).

However, the above trend (volcanism leading to increase in GHG causing global warming, and in the process threatening extinction of life) may also occur in a reverse way. Computer models confirm that when glaciers melt, they reduce the pressure on continents. Melting of ice, however, pushes the sea level to rise, thereby increasing in the process pressures on the ocean floor. This change in pressure levels on the Earth's crust seems to cause increases in volcanic eruptions. The speed of the transition from the Ice Age to melting, rather than the total amount of melting, may influence intensity of the volcanic eruptions (Kutterolf et al., 2013).

In contrast, isolated volcanism in the modern centuries may have little effect. For example, the annual CO_2 emitted by submarine and subaerial volcanoes has been estimated as 72–107 million tons and 267 million tons, respectively, which is however much less than that contributed by human activities (32 billion tons/year, Le Quere et al., 2009; Morner and Etiope, 2002). Interestingly, volcanic eruption could also introduce a cooling effect. Volcanoes emit sulfate aerosols which reflect back incoming sunlight, cooling the planet in the process. A large volcanic eruption such as the Pinatubo eruption in 1991, in fact, had a global cooling effect of 0.1–0.3°C for several years (Harris and Mann, 2014; Robock, 1994; Zielinski, 2000).

2.1.2 Anthropogenic Causes: Greenhouse Gas

The Earth's climate witnessed changes throughout its history. The last one completed about 7000 years ago, marking the beginning of the modern climate era and human civilization. Such alterations were affected by the GHG and aerosols by altering incoming solar radiation and outgoing infrared (thermal) radiation. The overall intensity of impact depends on three factors—abundance of particular GHG, residence time of such GHG, and its global warming potential

(GWP). For example, carbon dioxide (a member of GHG family) has high abundance in the atmosphere and resides for hundreds of years. In contrast, methane although stays only for 10 years in the atmosphere, it is about 21 times more efficient in absorbing radiation than CO_2, clocking a high GWP.

Hence, in contrast to the factors that contribute naturally to warming, and consequent modification of the global climate (see Section 2.1.1), a host of evidences are emerging to suggest that human activity is the primary reason for recent warming. These assessments show that the recent rise in temperature is clearly unusual in at least during the last 1000 years (see Fig. 1.3), when no major variation in solar energy input had been noticed (IPCC, 2007, 2014).

In fact, about 80% of such increase in GHG emission was believed to have come from the burning of fossil fuels, while the rest was contributed, among others, by deforestation and associated agricultural practices. In fact, the emission by the end of the 21st century is predicted to reach 900 ppm, almost three times more than that which occurred once during the glacial–interglacial era (IPCC, 2014). The concentration of CO_2 in the atmosphere is expected to alter the average global surface temperature (GST) over land and sea, in a similar way as it did during the 20th century. The GST shows an increase of approximately 0.78 °C since the year 1900. It may be noted that the 20 warmest years have all occurred since 1981, and the 10 warmest since 2004 (NCDC-NOAA, 2016, Table 1.1). Salient characteristics of major GHG are tabulated in Table 2.1, and briefly discussed below.

2.1.2.1 Carbon Dioxide

CO_2 is the primary greenhouse gas (GHG) and is naturally present in the atmosphere as part of the Earth's carbon cycle (cycle involves natural circulation of carbon through the atmosphere, oceans, soil, plants, and animals). CO_2 is contributed largely by plant respiration, volcanic eruptions, deforestation, combustion of fuel/oil/diesel, and ocean–atmosphere exchange.

It is believed that carbon dioxide has been building up in the Earth's atmosphere since the beginning of the industrial era in the mid-17th century. Human activities are transmuting the carbon cycle—both by adding more CO_2 to the atmosphere and by exploiting the ability of natural sinks, like forests and oceans, to remove CO_2 from the atmosphere. The industrial revolution emitted CO_2 through combustion of fossil fuels (coal, natural gas, and oil) for energy and transportation, although certain industrial processes and land-use changes also emit additional CO_2 (Hardy, 2003).

Atmospheric CO_2 concentrations have increased by more than 40% since pre-industrial times. This concentration was recorded beyond 400 ppm in 2013 (Butler, 2013). At the global scale, CO_2 increased from 15 million Gg in 1970 to 32 million Gg in 2008 (Fig. 2.1). While South Asia showed only marginal increase in carbon dioxide emission during this period, such emission

Table 2.1 Characteristics of Major Greenhouse Gases

Greenhouse Gases	Chemical Formulae	Anthropogenic Sources	Atmospheric Lifetime (Years)	GWP (100-Year Time Horizon)
Carbon dioxide	CO_2	Fossil-fuel combustion, land-use conversion, cement production	~100	1
Methane	CH_4	Fossil fuels, rice paddies, waste dumps	12	25
Nitrous oxide	N_2O	Fertilizer, industrial processes, combustion	114	298
Tropospheric ozone	O_3	Fossil fuel combustion, industrial emissions, chemical solvents	Hours/days	NA
CFC-12	CCL_2F_2	Liquid coolants, foams	100	10,900
HCFC-22	CCl_2F_2	Refrigerants	12	1,810
Sulfur hexaflouride	SF_6	Dielectric fluid	3,200	22,800

	Pre-1750 Troposphere Concentration (Parts per Billion)	Current Troposphere Concentration (Parts per Billion)
Carbon dioxide	280,000	388,500
Methane	700	1,870/1,748
Nitrous oxide	270	323/322
Tropospheric ozone	25	34
CFC-12	0	0.534/0.532
HCFC-22	0	0.218/0.194
Sulfur hexaflouride	0	0.00712/0.00673

The GWP indicates the warming effect of a greenhouse gas, while the atmospheric lifetime expresses the total effect of a specific greenhouse gas after taking into account global sink availability. The lifetime indicates how long the gas remains in the atmosphere and increased radiative forcing quantifies the contribution to additional heating over an area. The vast majority of emissions are carbon dioxide followed by methane and nitrous oxide. Lesser amounts of CFC-12, HCFC-22, perflouroethane, and sulfur hexafluoride are also emitted and their contribution to global warming is magnified by their high GWP, although their total contribution is small compared to the other gases.
Source: Modified after IPCC, 2014. Climate Change 2014: Impacts, Adaptation, and Vulnerability. Part A: Global and Sectoral Aspects. In: Field, C.B., Barros, V.R., Dokken, D.J., Mach, K.J., Mastrandrea, M.D., Bilir, T.E., Chatterjee, M., Ebi, K.L., Estrada,Y.O., Genova, R.C., Girma, B., Kissel, E.S., Levy, A.N., MacCracken, S., Mastrandrea, P.R., White, L.L. (Eds.), Contribution of Working Group II to the Fifth Assessment Report of the Intergovernmental Panel on Climate Change. Cambridge University Press, Cambridge, UK, New York, NY, USA, pp. 1132; UNFCCC, 2017. United Nations Framework Convention on Climate Change. In: Bonn Climate Change Conference. UN Climate Change Centre, May 2017.

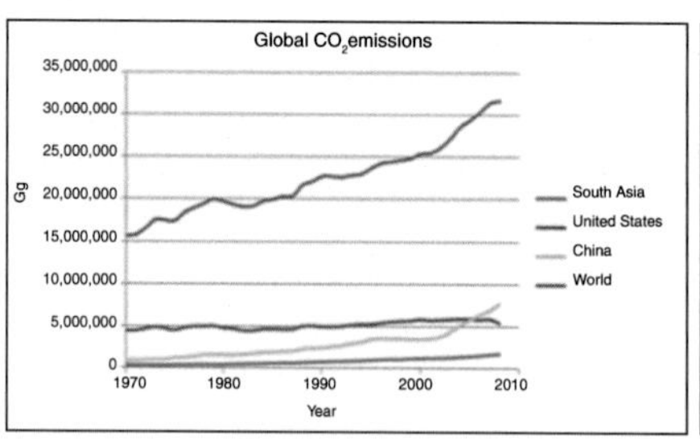

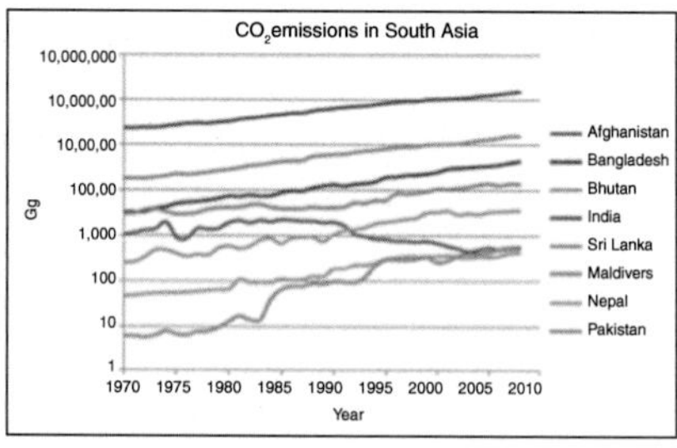

FIGURE 2.1 **Carbon dioxide emissions in South Asia and elsewhere in the world, since 1970.**
Source: Constructed from data of Ritchie and Roser (2017).

in USA has increased by about 5% between 1990 and 2012 (Langmuir and Broecker, 2012). However, China showed considerable increase from 1 million Gg to 7.5 million Gg between 1970 and 2008. All South Asian countries, except Afghanistan, showed variable increase in CO_2 emission from 1970 to 2008, with India taking the lead. While emission in India increased from 0.4 million Gg in 1970 to 1.2 million Gg in 2008, the emission amount in Afghanistan was actually reduced from 1000 Gg in 1970 to about 700 Gg in 2008. Interestingly, the maximum degree of increase in CO_2 emission has been shown by the tiny archipelago of Maldives—from 8 Gg in 1970 to about 800 Gg in 2008 (Ritchie and Roser, 2017).

2.1.2.2 Methane

CH_4 is the main component of natural gas and is used largely as fuel. In its natural state, methane is found both below ground and under the sea floor as gas hydrate. Methane is an important GHG with a GWP of 34 compared to CO_2 over a 100-year period (Table 2.1). The Earth's atmospheric methane accounts for 20% of the total radiative forcing from all of the long-lived and globally mixed GHG (Khalil, 1999). On average, about 55% to 65% of the emissions of atmospheric methane come from human activities (Dlugokencky et al., 2011). Since preindustrial times, methane levels have increased from 0.75 ppm by about 250% to their current levels of 1.85 ppm (Fig. 2.2).

The increase in levels of methane in the atmosphere has increased in recent decades primarily due to human activities. Agriculture plays a substantial role with—(a) rice producing methane via bacteria that live in the flooded fields, and (b) livestock producing methane in their digestive tracts. Moreover, mining coal, extraction and transport of natural gas, other fossil fuel-related activities, and waste disposal, including sewage and decomposing garbage in landfills, all play potent roles in methane emission (IPCC, 2014). It is found that one-ton emission of methane has the equivalent warming effect of little more than 32 tons of carbon dioxide. Thus it appears that methane emissions may have a bigger impact on global warming than previously thought (Etminan et al., 2016).

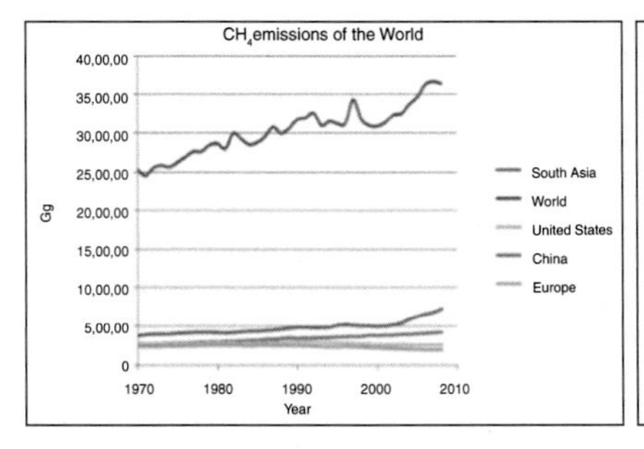

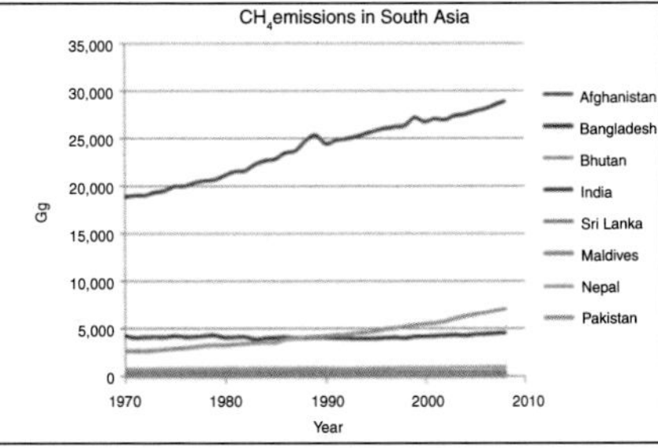

FIGURE 2.2 **Methane emissions in South Asia and elsewhere in the world, since 1970.** *Source: Constructed from data of Ritchie and Roser (2017).*

The emission of methane on the world scale shows an increase from 250,000 Gg in 1970 to more than 350,000 Gg in 2008 (Fig. 2.2). While USA and Europe show a downward trend in methane emission from about 25,000 Gg in 1970 to marginally less than that in 2008, such emission in South Asia and China have increased respectively from 25,000 Gg and 40,000 Gg in 1970 to 45,000 Gg and 70,000 Gg in 2008. Among the South Asian nations, India contributes maximum to the emission of methane (19,000 Gg in 1970 increased to 29,000 Gg in 2008), with Maldives the least. Pakistan (with 2500 Gg in 1970 and 7000 Gg in 2008) and Bangladesh (4000 Gg in 1970 increasing to 4900 Gg in 2008) take the second and third places in the subcontinent respectively for emitting methane. Sri Lanka comes fourth and almost maintained the emission level close to 1000 Gg between 1970 and 2008. The emission from the rest of the countries of South Asia has been below that of Sri Lanka (Fig. 2.2, Ritchie and Roser, 2017).

2.1.2.3 Nitrous Oxide

N_2O is a chemical compound and commonly known as laughing gas due to its ecstatic effects on inhaling. At room temperature, it is colorless, with a slightly sweet aroma and taste, and noninflammable in nature. Due to its euphoric property, it is used as an anesthetic/analgesic in surgery and dentistry. N_2O is also used as an oxidizer in the launching of satellite rockets and in car/bike racing to increase the power output of engines.

Nitrous oxide is naturally present in the atmosphere as part of the Earth's nitrogen cycle, although globally about 40% of total N_2O emissions come from human activities such as agriculture, fossil fuel combustion, wastewater management, and industrial processes (US-EPA, 2010). Nitrous oxide molecules stay in the atmosphere for an average of 114 years before being removed by a sink or destroyed through chemical reactions. The impact (GWP) of 1 pound of N_2O on warming the atmosphere is almost 298 times that of 1 pound of carbon dioxide. N_2O is also produced naturally in the soil during the microbial processes of both nitrification and denitrification (USGHGIR, 2011).

The levels of N_2O are increasing, primarily as a result of fertilizer use and fossil fuel burning. The concentration of nitrous oxide has increased by about 20% relative to preindustrial times. At the global scale, the N_2O emission increased from 7000 Gg in 1970 to 10,500 Gg in 2008 (Fig. 2.3). In contrast, N_2O emission in both China and South Asia, although started at 500 Gg in 1970, differed much mid way and ended by emitting in 2008 about 1900 Gg in China and little under 800 Gg in South Asia. The N_2O emission in USA has largely been static at 1000 Gg. In South Asia, India contributes a maximum of N_2O, ranging from 275 Gg in 1970 to 770 Gg in 2008. Rest of the South Asian countries contributes less than 100 Gg, with Afghanistan being the lowest (Ritchie and Roser, 2017).

2.1.2.4 Sulfur Dioxide

Coal-based thermal power plants are the main source from which sulfur dioxide (SO_2) is released. Since the beginning of the 21st century, there has been a steady rise in the SO_2 emissions from nearly 118,000 Gg in 1970 to about 125,000 Gg in 2008 (Fig. 2.4). In fact, the emission level in USA has dropped

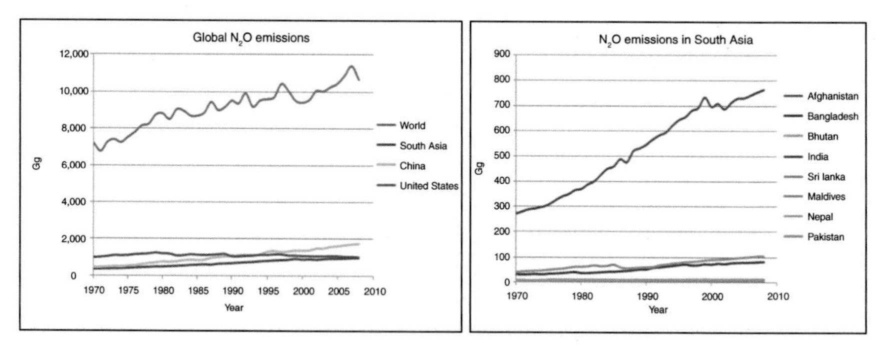

FIGURE 2.3 Nitrous oxide emissions in South Asia and elsewhere in the world, since 1970.
Source: Constructed from data of Ritchie and Roser (2017).

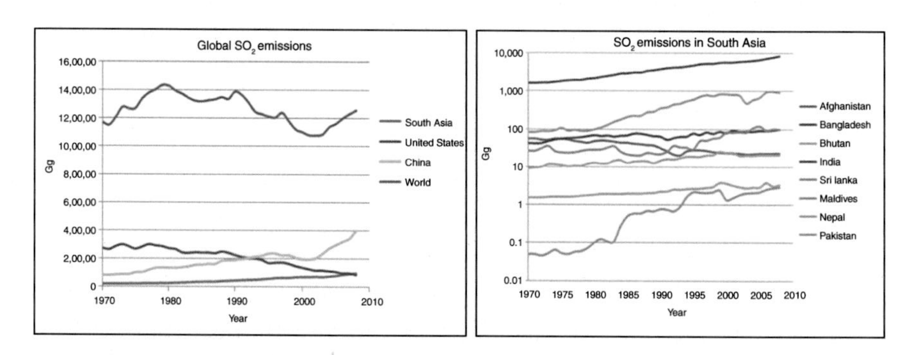

FIGURE 2.4 Sulfur dioxide emissions in South Asia and elsewhere in the world, since 1970.
Source: Constructed from data of Ritchie and Roser (2017).

down gradually since 1970, while that for China have increased. In South Asia, the rise in emissions has been slow and gradual. The country-wise distribution of SO_2 emissions in South Asia shows a gradual rise in India and Pakistan, while other South Asian nations show negligible increase, and remaining below 100 Gg since 1970 (Ritchie and Roser, 2017).

Sulfur dioxide emissions have substantial impacts on human health, as well as terrestrial and aquatic ecosystems, and have come under increasing regulation worldwide (Klimont et al., 2013). It is also a principal precursor of anthropogenic aerosols in the atmosphere, and acts at times as a cooling agent.

2.1.2.5 Chlorofluorocarbons

The halogenated paraffin hydrocarbon combination of carbon, chlorine, and fluorine *is known as Chlorofluorocarbons* (CFC). These are produced as volatile derivatives of methane, ethane, and propane, and are also known as Freon. The CFC are a member of the fluorocarbon and fluoroethane family that shows high GWP at times much higher even than that of CO_2 (Lu, 2013; Table 2.1). CFC are widely used in cooling, refrigeration, propellants (in aerosol applications), and solvents.

A comparative assessment of CFC emission during the last decade suggests that in Russia CFC emission reduced marginally from 1600 tons in 2005 to just above 1500 tons in 2009. During the same period, Japan and Germany reduced their CFC emissions respectively from 1400 and 900 tons to 1200 and 800 tons. Contrary to this trend, CFC emission in India increased from 1150 tons in 2005 to 1500 tons during the next four years.

CFC are blamed for the depletion of "good ozone" in the upper atmosphere (stratosphere 9–48 km above ground), which protects the Earth from harmful levels of ultraviolet radiation from the Sun. In addition, CFC also contributes in increasing the "bad ozone," occurring in the lower atmosphere (troposphere) that traps the heat (Ramanathan and Feng, 2009). Since the late 1800s, the average levels of ozone in the lower atmosphere (bad ozone) have increased by more than 30% (Lamarque et al., 2005).

Sulfur hexafluoride (SF_6) is an inorganic, colorless, odorless, nonflammable, but extremely potent GHG. With its density (6.12 g/L) considerably higher than the air (1.225 g/L), SF_6 has a far more GWP (23,900 times) than that of CO_2 (IPCC, 2007). It is an extremely stable, inert, long-lived GHG, having an estimated atmospheric lifetime of 800–3200 years (lifetime of CO_2 is only ~100 years). Due to its high relative density, it accumulates in low-lying areas, and could cause asphyxiation at high concentrations (Shriver and Atkins, 2010). The emission of SF_6 in the world since 1970 showed a steep increase from 0.9 Gg in 1970 to 4.9 in 2000 and 6.5 Gg in 2010, commensuration to increase in China, which showed an estimated increase from almost zero in 2000 to

1.9 Gg in 2008. Comparatively, India generates negligibly; but if compared with other South Asian countries, she remained the maximum contributor of SF_6, followed closely by Pakistan (Ritchie and Roser, 2017).

Hexafluoroethane (C_2F_6) is a fluorocarbon counterpart to the hydrocarbon ethane. It is a nonflammable, highly inert gas, and thus acts as an extremely stable GHG, with an atmospheric lifetime of 10,000 years and a GWP of 9200. Its radiative forcing is 0.001 W/m^2 (Bozin and Goodyear, 1968). These are emitted to the atmosphere mainly by the aluminium and semiconductor industries. Global emissions of this variety of perfluorocarbons calculated from atmospheric measurements are significantly greater than expected from reported national and industry-based emission inventories (Kim et al., 2014). The C_2F_6 emissions in the world varied from 1.25 Gg in 1970 to 2.5 Gg in 2010. A gradual increase in emission of C_2F_6 is seen in China, from zero in 1990 to 0.5 Gg by about 2010. The highest emission of C_2F_6 in India has been recorded in 1991 (0.020 Gg) and the least in 1974 (0.009 Gg) and 2005 (0.012 Gg; IPCC, 2014; Ritchie and Roser, 2017).

Perfluorocarbons (PFC) are chemically inert and nontoxic gases. With low boiling property and long atmospheric lifetime, PFC remains a potent GHG. The emission of global PFC shows an increase from 85,000 Gg in 1970 to 100,000 Gg in 2010, after dipping from 120,000 Gg around 1990. Effective measures brought PFC emission in USA down from 30,000 Gg in 1970 to about 10,000 Gg in 2010. However, such emission in Canada (~8000 Gg) and in India (~2000 Gg) almost remained the same during the four decades that ended in 2010. In India, PFC emission fluctuated between 1020 Gg in 1974 and 2350 Gg in 1991 (Ritchie and Roser, 2017).

Tetrafluoromethane (CF_4) is caused generally from primary aluminium production and acts as a GHG. The GWP of CF4 is 6500 and is one of the significant GHG emitted by the petrochemical industry (Benchiata, 2013). The concentration of CF_4 emission at the global scale did not change much between 1970 and 2010 (remains at 11 Gg), except recording 17 Gg in 1978 and 15 Gg in 1990. While the emission of CF_4 in USA is reduced from 4 Gg in 1970 to almost 0.70 Gg in 2010, the same in China increased from negligible in 1970 to 2 Gg in 2010. Emission in India, however, remained very low during the entire four decades (1970–2010; 0.10–0.20 Gg) (Ritchie and Roser, 2017).

Nitrogen trifluoride (NF_3) is a chemical that is released in some high-tech industries, including in the manufacturing of electronic gadgets- semi-conductors and LCD (Liquid Crystal Display) panels, and certain types of solar panels and chemical lasers. The emission of NF_3 is increasing rapidly as industrial production increases and is particularly alarming because NF_3 can reside in the atmosphere for 100 years with a GWP of 17,200 (Russell, 2013). The emission of NF_3 at the global scale increased from 0 Gg in 1990 to 0.175 Gg in 2008. Emission of NF_3 in USA increased from almost negligible amount in the year 2000 to 0.028 Gg in 2008. In India, the NF_3 emission fluctuated from less than 0.0002 in 2001 to 0.001 in 2004. The emission then reduced to negligible amounts in 2005, which, however, again rose beyond 0.001 Gg in 2008. The global emissions of NF_3 in 2011 were 1.18 ± 0.21 Gg (Ritchie and Roser, 2017).

Hydrofluorocarbons (HFC) are among the atmosphere's fastest growing, and most potent GHG (Lunt et al., 2015). Its emission at global scale increased from about 25,000 Gg in 1970 to 650,000 Gg in 2008, and the main contributor had been the USA and China (increased from 0 to 250,000 Gg and 0 to 150,000 Gg, respectively). Europe also contributed to the overall increase of HFC emission in the world with 90,000 Gg in 2008, while contribution of India and Canada remained very low.

2.1.2.6 Water Vapor

It is the gaseous phase of water and constitutes an important GHG. It can be produced from evaporation on boiling of water/liquid and from sublimation of ice. It is lighter than air and setoff convection currents that can lead to formation of clouds. Various studies suggest that the clouds and water vapor contribute about 66% to 75% of the total global warming effect, while CO_2 contributes only 20% (Kiehl and Trenberth, 1997). Because water vapor is a positive feedback GHG, a 1°C change caused by extra CO_2 from fossil fuels, the water vapor will cause the temperature to go up another 1°C. The higher concentration of water vapor in the atmosphere, may able to absorb more thermal infrared energy radiated from the Earth, thus further warming the atmosphere. The warmer atmosphere can then hold more water vapor, and thus the process repeats (Frank, 2015).

Sector-wise emission details of GHG in South Asian countries provide interesting information (Table 2.2). The sectors include tourism, power generation, transportation, industrial and domestic sectors, and agriculture. For example, the emission of carbon dioxide from various activity sectors in the case of Sri Lanka are particularly shown in Fig. 2.5. As a stand-alone example, the emission of GHG and aerosols in South Asia in the year 2008 is shown in Table 2.2.

2.1.3 Anthropogenic Causes: Aerosols

Coupled with GHG, aerosols in the atmosphere also influence solar radiation balance of the Earth. An aerosol is a colloid of fine suspended solid particles or liquid droplets, in air or in gas. Aerosols can be natural, such as (a) black carbon or inorganic carbon comprising volcanic eruptives, desert dust,

Table 2.2 Sector-Wise GHG Emissions (%) in South Asian Countries

Sectors	AFG	BNG	BHU	IND	MLD	NPL	PAK	SLN
Tourism + Commercial	-	-	-	10	36	06	-	-
Energy & Electricity	-	33	17	-	27	-	46	65
Transportation	-	-	-	24	23	49	-	-
Residence (+ LUCF)	-	17	-	28	-	12	03	-
Industry	01	02	15	33	-	25	05	02
Others (Waste + Fishery)	22	10	03	05	14	-	03	10
Agriculture	42	39	64	-	-	08	43	23

LUCF= Land Use Change & Forestry; Compiled from Marcu et al (2015), Lohani and Baral (2011), Sridhar (2010), USAID; WRI CAIT 2.0 Database (2015); FAO STAT (2015), Ranasinghe (2010), TNA-2016.

Source: Compiled from Marcu, A., Stoefs, W., Belis, D., Katja, T., 2015. Country Case Study—Maldives Climate for Sustainable Growth. Center for European Policy Studies (CEPS), Brussels, p. 51; Lohani SP, Baral B., 2012. Conceptual framework of low carbon strategy for Nepal. Low Carbon Econ. 2: 230–238; Sridhar, K.S., 2010. Carbon emissions, climate change and impacts in India's cities, in India Infrastructure Report 2010: Infrastructure Development in a Sustainable Low Carbon Economy: Road Ahead for India, (eds., 3iNetwork), pp. 345–354, Oxford University Press: New Delhi; US AID, 2015. United States AID Agency Financial Report, 168p; WRI CAIT 2.0 Database (2015); FAO, 2015. Climate Change and Food Systems: Global Assessment for Food Security and Trade. Food and Agricultural Organization of the United Nations, Rome, pp. 1–356; Ranasinghe, D.M.K.H. Climate change mitigation-Sri Lanka's perspective. Proceedings of the 15th International Forestry and Environment Symposium, Nov 26–27, 2010. Published by the Department of Forestry and Environmental Science, University of Sri Jayewardenapura, Sri Lanka, pp. 290–296, 2010; Technology Needs Assessment (TNA) Report. Climate change mitigation. Govt. of Pakistan, Ministry of Climate change, Islamabad, Pakistan. Funded by Global Environment Facility (GEF) and implemented by UNEP and UNEP DTU Partnership in collaboration with Regional Center, Asian Institute of Technology, Bangkok, pp. 1–112, 2016.

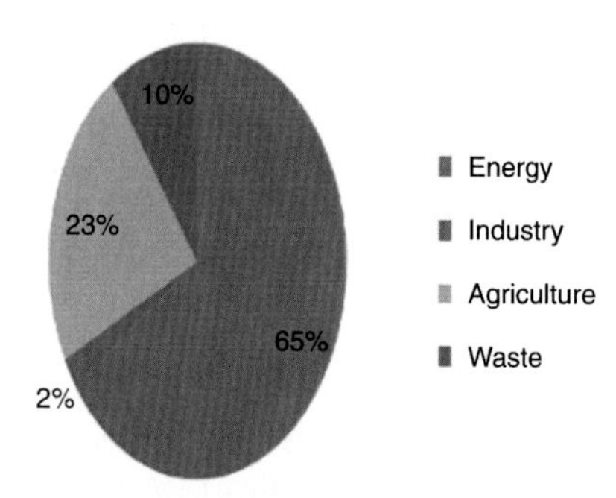

FIGURE 2.5 Percentage contribution from different sectors in GHG emossions in Sri Lanka. *(After Meisen, P., Azizy, P., 2008. Rural Electrification in Afghanistan, How Do We Electrify the Villages of Afghanistan? GENI (Global Energy Network Institute), California, USA, pp. 1–26; Ranasinghe (2010)).*

sea salt, water droplets, sulfuric acid droplets, fog, plant exudates, and geyser steam, (b) organic carbon comprising smoke, pollen, spores, bacteria, forest exudates; and (c) anthropogenic or artificial (e.g., haze, ashes, smoke, deodorant sprays, soot, and fumes). When these minute particles become sufficiently large, we notice their presence as they scatter and absorb sunlight. The aerosols can affect the climate by warming or cooling the Earth (Hinds, 1999). These aerosol particles are so minute (<10 micrometer = μm) that they can get into the lungs, potentially causing serious health hazards. In comparison, a human hair-top is about 100 micrometers, so roughly 10 or more aerosol particles could be placed on each hair.

Organic carbon (OC) is the finest variety of aerosol (size between 0.7 and 0.22 μm, much smaller than $PM_{2.5}$) and carbonaceous in composition. High amounts of organic matter are common in low oxygen areas, such as bogs and wetlands. Globally, the OC increased from little more than 4000 Gg/year in 1850 to close to 9500 Gg/year in the year 2000. This offers an increase of about 45% during 150 years (average increase 35 Gg annually, Sloss, 2012).

There has always been an allegation that South Asia has played a hazardous role toward global warming. To test the validity of this statement, the values of OC and BC emission by South Asia in comparison to the rest of the world were plotted. Between 1850 and 2000, while contribution of BC and OC in South Asia varied from 100 to 600 Gg/year and 200 to 1500 Gg/year respectively, the same for the world varied from 100 to >4500 and >4000 to ~9500 Gg/year, respectively. Hence the allegation appears to be misplaced (Fig. 2.6 , Ritchie and Roser, 2017).

Black carbon (BC) is a fine particulate matter (0.7 to 2.5 μm, $PM_{2.5}$) and is normally formed through incomplete combustion of fossil fuel, biofuel and biomass (Anenberg et al., 2012). Black carbon warms the Earth by absorbing heat

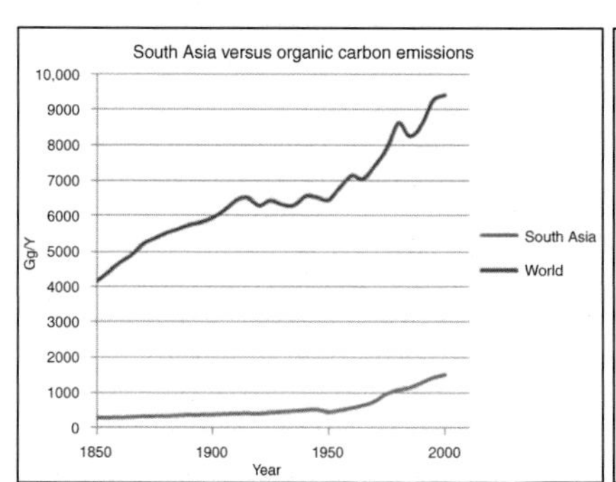

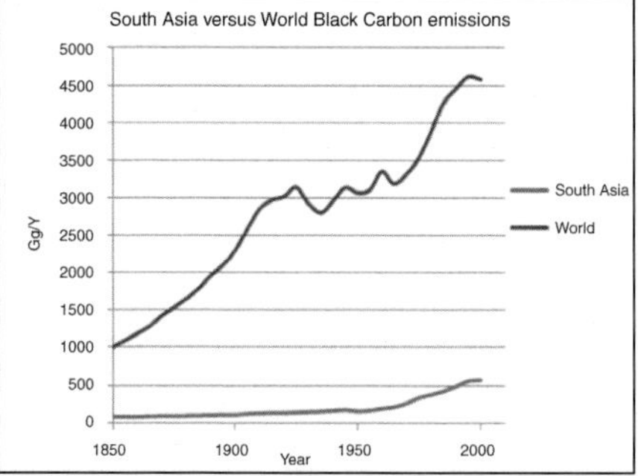

FIGURE 2.6 Concentrations of organic carbon and black carbon in South Asia and elsewhere in the world, since 1970. *Source: Constructed from data of Ritchie and Roser (2017).*

in the atmosphere and by reducing albedo effect (ability to reflect sunlight) when deposited on snow and ice. BC significantly contributes to the fertility of soil as it absorbs important plant nutrients (Glaser, 2007; Ramanathan and Carmichael, 2008).

Black Carbon is a short-lived climate pollutant (residence time only few weeks) and could have a rapid impact, especially at local scale (Hallegatte et al., 2016; Rogelj et al., 2014). Hence, reducing BC emission could quickly slow down the rate of climate change. However, globally, BC increased from 1000 Gg/year in 1850 to close to 4600 Gg/year in the year 2000, an increase by 22% (average annual increase 24 Gg/year, Sloss, 2012). Asia, including China and India, accounts for 40% of the entire global BC emissions. In South Asia, India ranks highest with 64% emissions, followed by Pakistan (22%), Bangladesh (8%), Nepal (4%), Sri Lanka (2%), and Bhutan (<1%). It seems, burning of fossil fuel, use of chullah (open-air cooking) in villages, and residential biofuel combustion account for 25%, 33%, and 42% of BC emissions in India, respectively (UNEPDA, 2014; Fig. 2.6).

The larger size of particulate aerosol matter (size rage 2.5 to 10 μm, PM_{10}) shows a variable trend since 1970 and displays stark difference between the world and South Asia. The PM_{10} emission in the world has been fluctuating from 75,000 Gg to 135,000 Gg between 1970 and 2010. In contrast, the slow, gentle, and gradual increase in PM_{10} emission in South Asia from 5000 Gg in 1970 to about 15,000 Gg in 2010 could also be seen in a countrywise distribution. India remains the highest contributor of PM_{10} emission in South Asia (~7000 Gg in 1970 to more than 10,000 Gg in 2010), and Maldives has been the least (less than 0.1 to 0.5 Gg) during the same time period (Fig. 2.7 , Ritchie and Roser, 2017).

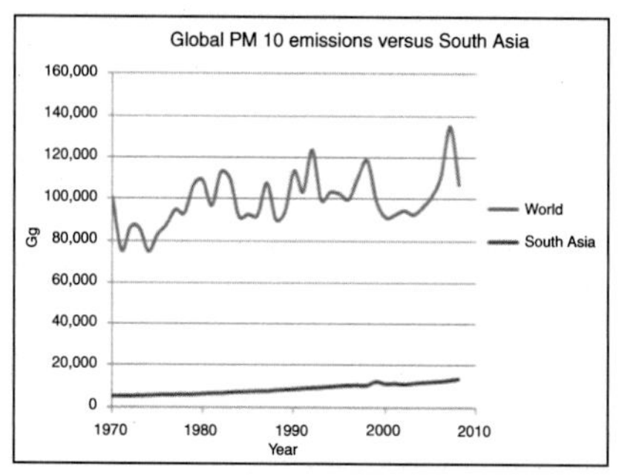

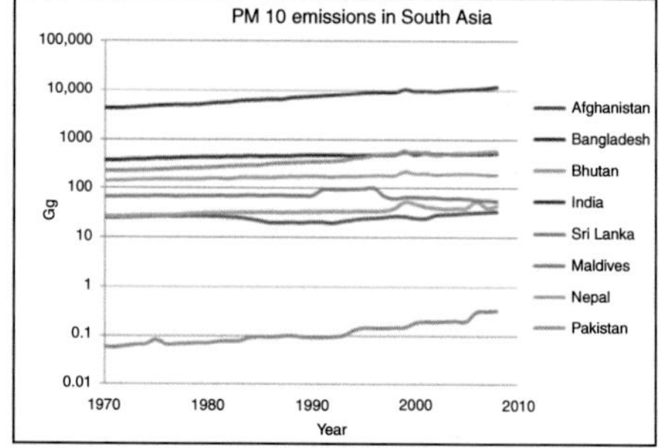

FIGURE 2.7 Concentrations of particulate matter in South Asia and elsewhere in the world, since 1970. *Source: Constructed from data of Ritchie and Roser (2017).*

2.1.4 Anthropogenic Causes: Terrestrial

2.1.4.1 Deforestation

Climate has also been affected by a range of human activities. These activities include felling of trees and burning of forests, replacing natural vegetation with agriculture or urbanization. These actions also impact large-scale irrigation, forest management, mining, infrastructure projects, and fire hazards. Deforestation can also alter the amount of heat reflected or absorbed by the Earth's surface, causing local and even regional cooling or warming.

The net carbon uptake by terrestrial forestry is likely to increase by 2050 amplifying climate change. It is estimated that increase in global average temperature by 1.5°C to 2.5°C will cause extinction of about 20 to 30% of plant and animal species, thus endangering the biodiversity (IPCC, 2014). For example, India is one of the 12 mega biodiversity countries of the world and encompasses about 11 per cent of the world's flora, which will be affected dearly (Chitale et al., 2014). Several strategies are being formulated (e.g., REDD + = *reducing emissions from deforestation and forest degradation*) by various nations to provide incentives to tropical nations and states that reduce their GHG emissions from deforestation and forest degradation.

2.1.4.2 Livestock Management

Among others, meat consumption plays a significant role in aggravating global warming and climate change. It is estimated that about 15% of all emissions come from worldwide livestock. Globally, meat consumption has witnessed an exponential growth from 70 million tons in 1961 to 278 million tons in 2009—a whopping 300% increase. Meat consumption is expected to rise further to 460 million tons by 2050. All this meat, the bulk of which is beef (contributing alone over 40% of emissions), is grown in large industrial farms in Europe, United States, South America, and Australia. A kilogram of beef emits 22.6 kg CO_2, compared to 2.5 kg from pork, 1.6 kg from poultry, 1.3 kg from milk, and 0.8 kg from wheat production (FAO, 2015).

The farm animals themselves emanate an extensive amount of methane, which is generated in their digestive system. In 2013, such emission through enteric fermentation generated 39% of all livestock emissions. More animals (cattles, pigs, etc.) mean more forestland is used to grow fodder for the livestock. In fact, almost a quarter of the world's cultivable land is used for this purpose, and the fertilizers used contribute to nitrogen oxide emissions. The loss of forests contributes another 21% of emissions. The manure produced by these animals runs into millions of tons and produces 26% of emissions, mostly nitrogen oxides and methane. The remaining 14% are generated from various energy usage such as for transportation, processing and from land-use changes—like deforestation—to provide land for growing feed (FAO, 2015; Rojas-Downing et al., 2017). The average meat consumption in USA is about 322 grams/day/person, with Australia and New Zealand consuming close behind. Europeans, Brazilians, Argentines, and Venezuelans eat about 200 grams per day on an average, while the Chinese consume 160 grams. Indians, however, consume just 12 grams/day/person, whereas global average is 115 grams/person/day (FAO, 2015; Rojas-Downing et al., 2017). Hence, minimizing meat consumption (mainly beef) can be highly beneficial in fighting against climate change, provided the big meat-eating countries take it up. Following the adage "Charity begins at home," we suggest that the world could start a "2-day beef-less" week.

2.2 IMPACT OF CLIMATE CHANGE

2.2.1 Impact on Atmosphere and Ocean

Climate is closely related to Geography. The formation of the mighty Himalayas as a result of India-Eurasia collision blocked the frigid Central Asian air, preventing it from blowing down south. This made the climate of South Asia significantly warmer and more tropical in character than it would otherwise have been (Wolpert, 1999). Again, for example, earlier in the Holocene Epoch (4800–6300 years ago), parts of what is now the Thar Desert were saturated enough to support perennial lakes. This phenomenon led researchers to propose that it was due to a much higher winter precipitation that coincided with stronger monsoons (Enzel et al., 1999). Similarly, Kashmir, which once had a warm subtropical climate, shifted to a substantially cooler temperate climate around 2.6–3.7 million years ago; and is repeatedly subjected to extended cold spells since the last 1 Ma (Pant, 2003). The areas of atmosphere and ocean, which are expected to be impacted by climate change, are accounted below.

2.2.1.1 Temperature

Due to high degree of accumulation of GHG in the atmosphere (see earlier Section 2.1), the average surface temperature of the Earth rose by $0.74 \pm 0.18\,^{\circ}C$ over the period between 1906 and 2005 (IPCC, 2013). More accurately, the world had warmed by $0.13 \pm 0.03\,^{\circ}C$ per decade between 1956 and 2006, compared to $0.07 \pm 0.02\,^{\circ}C$ increase per decade before 1956. Jansen et al. (2007) found that the temperature was relatively stable before 1850. While 1998 has been the warmest year (thanks to El Nino, see Section 2.1.1) followed by the years 2005, 2003, and 2010, the coolest year has been the 2011 (due to reverse-effect of La Nina, Langmuir and Broecker, 2012).

Annual combined land–surface–air–sea temperature at global and hemispheric levels from 1850 to 2006 shows that while temperature anomaly on land varied between $-0.6\,^{\circ}C$ and $1.1\,^{\circ}C$ during the period 1880 to 2010, the inconsistency in sea was much less between $-0.45\,^{\circ}C$ and $0.50\,^{\circ}C$ (Table 1.1).

The fluctuations/anomalies are found more on land. The base period 1901–2000 shows the month of July as the central median with maximum temperature (Brohan et al., 2006; NOAA, 2016; Sutton et al., 2007).

Evidence for changes in the climate system is nearly omnipresent—from the top of the atmosphere to the depths of the oceans (Kennedy et al., 2010). Temperatures at the surface, in the troposphere (the active weather layer extending up to about 8 to 16 km above the ground), and in the oceans have all increased over recent decades. The largest increase in temperature occurs closer to the poles, especially in the Arctic, where snow and ice cover have decreased substantially (Karl et al., 2009; Tomkins and Deconscini, 2015; Walsh et al., 2014a,b).

The average global temperature since Precambrian times offers interesting food for thought (Scotese, 2015; Fig. 1.2). The average temperature varied between 12°C (cool) and 22°C (warm), passing through 17°C (dry). It is found that even when there was no contribution from anthropogenic activities, the temperature fluctuated on several occasions between warm and cool in the geological past. For example, the average temperature was cool (~12°C) on four occasions—during Precambrian, at the boundaries between Ordovician and Silurian, and again between Carboniferous and Permian, and at the present time. The average temperature was however, warm (~22°C) and was found at the boundaries between Cambrian and Ordovician, Silurian and Devonian, Triassic and Jurassic, and mid Cretaceous to mid Tertiary.

The concentration of CO_2 in the atmosphere over the last 400,000 years appears to display almost similar trend with increase in temperature (Petit et al., 1999). A total of 5 peaks reaching to increase in 2°C in the atmosphere have been recorded at 410,000 years, 320–330,000 years, 230–240,000 years, 120–130,000 years, and today. Correspondingly, at the same period, the global temperatures had also increased to 280 ppmv, 300 ppmv, 280 ppmv, 290 ppmv, and 285 ppmv, respectively (ppmv= parts per million by volume =1 mL/L).

2.2.1.2 Cyclones and Storms

With over 600 million people living in low-lying coastal areas of the world (McGranahan et al., 2007), coastal floods can have devastating societal impacts. It is estimated that on an average, globally, 0.8–1.1 million people per year are affected by flood (Hinkel et al., 2014). Rise in temperature leads to more evaporation causing increased rainfall and snowfall.

The data with regard to frequency of occurrence of calamitous events (cyclones and storms) in the world oceans since 1945 has been collated and integrated (Fig. 2.8). The year 1975 recorded maximum number of cyclones and storms (20 times) in the northern Indian Ocean, which also separates a strikingly high annual average occurrence of extreme events between 1945 and 1975 (av. 18) and low annual average of occurrence after 1975 till 2010 (av. 5). In contrast, the southern Indian Ocean records high average occurrence of extreme events

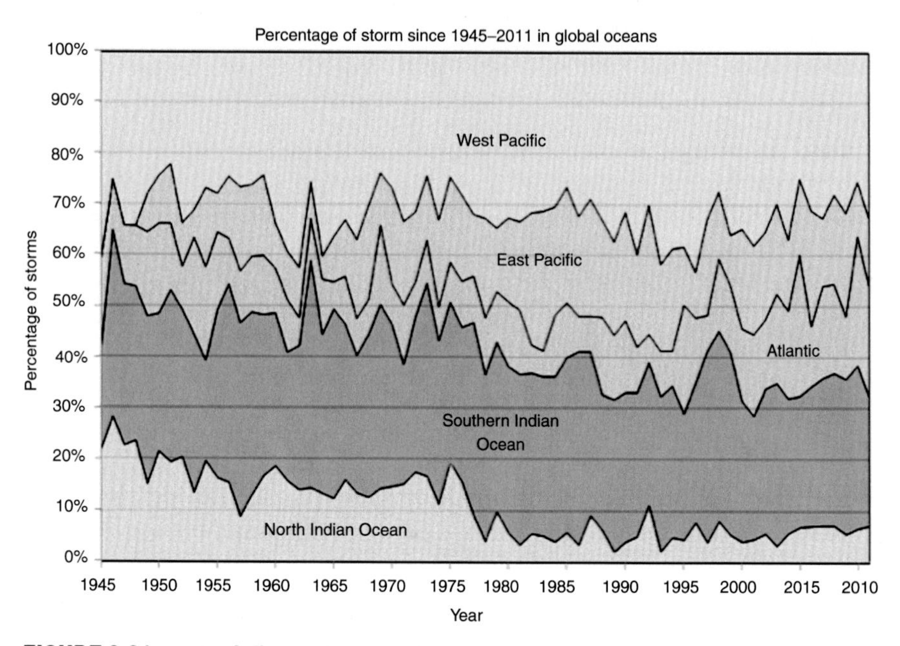

FIGURE 2.8 Impacts of climate change on South Asia: extreme events.
Source: Modified after IPCC (2014), UNFCCC (2017).

per year (excess of 20), with maximum of 49 events occurring in 1963 and lowest 12 events in the year 1945 (Jansen et al., 2007).

Similarly, the annual occurrence of cyclones and storms in the eastern Pacific has been low (less than 10 events/year), with the highest number of 27 events occurring in the year 1993 and the lowest number of 4 events in 1954. The western Pacific, in contrast, recorded high occurrence of extreme events (average about 24 events/year), with the highest number of 45 events in 1964 and the lowest number of 15 events in 1946 (Fig. 2.8). In the Atlantic Ocean, the highest number of extreme events (28) occurred in 2006, while the least number of events (04) occurred in 1983. The average number of such events in the Atlantic Ocean between 1945 and 2010 has been around 8 events per year. Percentage wise, the storms and cyclones in the world oceans since 1945 have been highest in the west Pacific followed by the east Pacific, southern Indian Ocean, Atlantic Ocean, and northern Indian Ocean (Muis et al., 2016; Rappaport, 2014; Resio and Westerink, 2008).

In the Indian Ocean, tropical cyclones form mostly in the Bay of Bengal compared to the Arabian Sea in 4:1 ratio and move in a WNW direction. In the past 270 years, 20 out of 23 major cyclone disasters (with loss of ≥10,000 lives) that occurred globally have swept the coastal regions of India and Bangladesh (Table 2.3). The world's highest recorded storm tide was about 12.5 m (about 41 ft), observed in association with the 1876 Backerganj cyclone near Meghna

Table 2.3A Major Disasters in South Asia

Sl	Type of Disaster	Number of Disasters		Loss of Life Numbers	
		1985–97	1998–2009	1985–97	1998–2009
1	Droughts	6	9	300	200
2	Earthquakes	27	35	13,166	155,233
3	Extreme temperature	28	41	4,011	10,021
4	Floods	141	235	32,610	26,111
5	Storms	97	99	165,360	23,056

Total number of disasters in South Asia: 1985–89: 96, 1990–94: 135, 1995–99: 134, 2000–2004: 166, 2005–2009: 199.
Total cost of the property lost during natural calamities and disasters: US$44,787,984,000.
Constructed by US based on data from Gaiha et al (2010) and SAARC Web-portal (www.saarc-sec.org) Gaiha R, Hill K, Thapa G 2010. Natural Disasters in South Asia, ASARC Working paper, 2010/06, In: Routledge's Handbook of South Asian economics, 1-25

Table 2.3B Number of Climate Disasters in South Asia between 2005 and 2014

Country	Number of Flood Events	Number of Drought Events
Afghanistan	41	3
Bangladesh	16	1
India	93	1
Maldives	1	–
Nepal	14	2
Pakistan	35	–
Sri Lanka	20	1

Source: *EM-DAT: The Emergency Events Database –Universitécatholique de Louvain (UCL) –CRED, D. Guha-Sapir –www.emdat.be, Brussels, Belgium.*

Estuary in Bangladesh. In October 1737, a storm tide height of 12.1 m was observed in West Bengal in the mouth of Hooghly River. The Orissa super cyclone of October 1999 that generated a wind speed of 252 km/hour with an ensuing surge of 7–9 m at Paradip caused unmatched inland inundation to the extent of 35 km from the coast. The devastation in all these cases was escalated more due to low and flat coastal terrain, high density of population, low awareness among community, inadequate preparedness, and absence of standard operating procedure (World Bank, 2012).

Tsunamis are oceanic waves of extremely long wavelength of the order of about 100–250 km caused by volcanism, landslides, and earthquakes occurring below the ocean floor. South Asia has two tsunamigenic zones, Andaman-Sumatra trench in the Bay of Bengal and the Makran coast in the Arabian Sea. The 2004 Indian Ocean tsunami that hit Indonesia, Sri Lanka, India, Thailand, and the Maldives took away about 230,000 lives and inflicted colossal damage to property, infrastructure, and economy.

2.2.1.3 Rainfall/Precipitation

Climate change can influence water resources and freshwater ecosystem, causing (a) drought through intense evapo-transpiration of water from soil, plants, and water bodies, and (b) cloudburst, as a warmer atmosphere can hold more water vapor and increase risk of intense localized rainfall. In fact, the atmosphere now holds 4% more water vapor than it did 40 years ago as a result of increasing temperatures. Changes in SSTs also play an important role in the atmospheric circulation and precipitation, causing droughts and abnormal precipitation, especially in tropics like South Asia.

IPCC (2013) data finds that heavy rainfall events have increased both in frequency and in their intensity by 20%. Hurricanes in the Atlantic and eastern Pacific (and following rainfall) have been severe over the last few decades. Australia and Southeast Asia have experienced heavy rainfall and increased droughts, and similar patterns are being observed worldwide, coinciding with rising temperatures over the past 50 years. Heat waves are also occurring more frequently as temperature shift upwards (USEPA, 2016; NOAA, 2016).

There is an increase in rainfall from 1901 (~1100 mm) to 2011 (nearly 1380 mm). Although the pattern of rainfall in South Asia had been fluctuating, a subtle increase in rainy days in a year has been noticed during those years. For example, in 1916, 1450 mm of rainfall was recorded in 170 days, while 150 days of rainfall had collected 1400 mm in 1956. However, though 1000 mm of rainfall was recorded in 100 days in 1966, the year 2003 recorded 1020 mm of rainfall in 125 days (i.e., only 20 mm of rainfall was recorded in additional 25 days). Hence, it can safely be said that duration of rain does not necessarily proportionately increase the rainfall amount (IMD, 2012).

The major floods occurring in various countries in South Asia during 1985 and 2010 give interesting results. The number of floods persistently increased from a rare 5 in 1985 to 31 in 2010 (Figs. 2.9 and 2.10). The annual number of floods was found to vary from merely 1 in 1986 to as many as 36 in 2007. The number of floods averaged for 5 years, showing stepwise increase from 11 in 1985–90 to 15 in 1991–95, 14 in 1995–2000, 22 in 2001–05 and ultimately to 26 in the period between 2006 and 2010. A related study to find out the reasons for such a heavy incessant rainfall in South Asia finds local, sudden torrential rain (including cloudburst) causing 46% of floods. The next most causal factor for floods in South Asia had been monsoon (41%), while precipitation associated with cyclones caused floods in 6% of the cases. Melting of snow and associated rain caused 3% of floods in South Asia, while the rest of 4% floods were caused by undefined factors (Fig. 2.10; IMD, 2012).

Even the two wettest places in the world—Cherrapunji and Mawsynram (both located in the northeastern part of India in the state of Meghalaya) have recorded erratic and sparse rainfall intervened by long dry spell in the recent

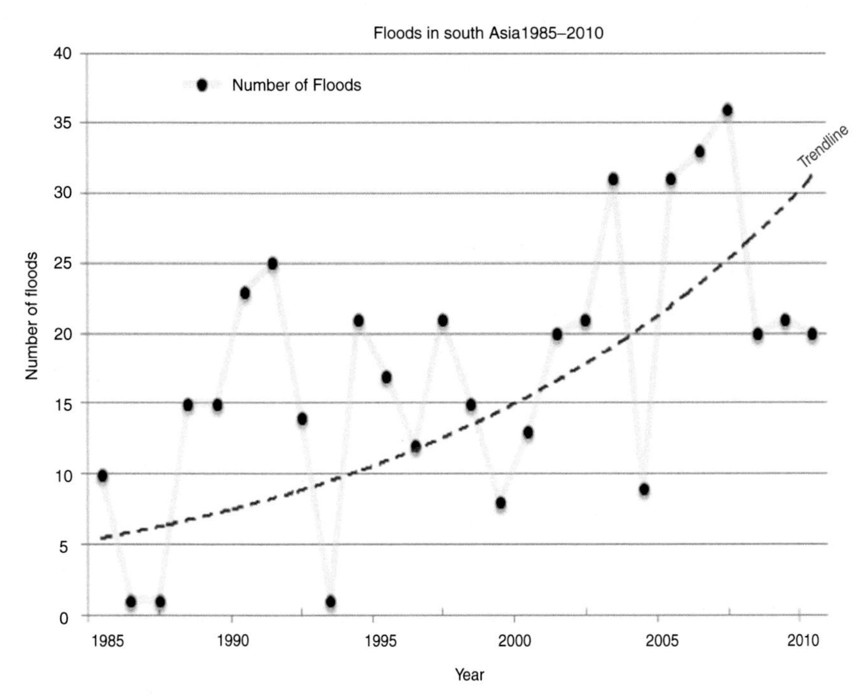

FIGURE 2.9 **Impacts of climate change on South Asia: floods.**
Source: Modified after IPCC, 2014; UNFCCC, 2017.

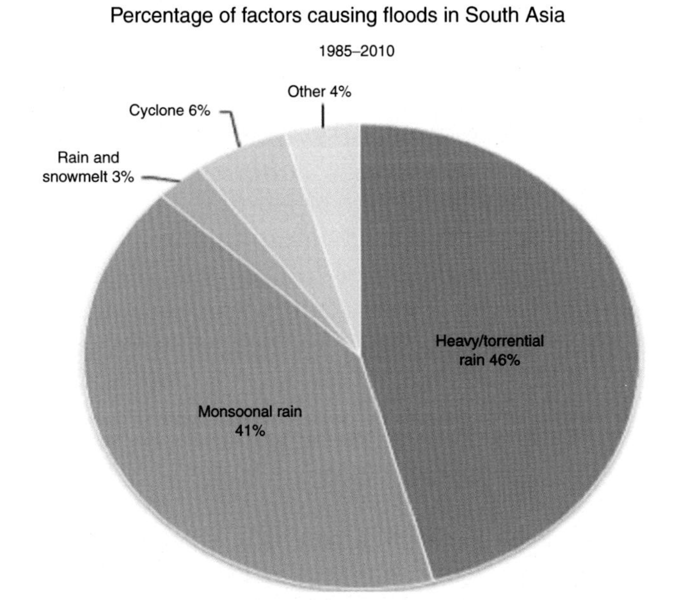

FIGURE 2.10 **Major factors causing floods in South Asia.**
Source: Modified after IPCC, 2014; UNFCCC, 2017.

past (Sethi, 2007). The water shortage is affecting the livelihoods of thousands of villagers who cultivate paddy and maize, despite these areas being densely covered by vegetation.

2.2.1.4 Melting of Glacier

Approximately 160,000 glaciers are found on the Earth, at virtually all latitudes from the tropics to the poles (except in Australia). The volume of ice in a glacier and correspondingly its surface area, thickness, and length is determined by the balance between inputs (accumulation of snow and ice) and outputs (melting and calving). And the balance between inputs and outputs is influenced by temperature, precipitation, humidity, wind speed, and other factors such as slope and the reflectivity of the glacier surface—all affecting (Fitzharris., 1996; IPCC, 2014; NCDC-NOAA, 2016).

Retreat of glacier is reported from the European Alps (by 30%–40%), New Zealand (by 25%), Mt. Kenya, and Kilimanjaro (by 60%), and by much faster rate in the Peruvian Andes (Chinn, 1996; Haeberli and Beniston, 1998; Hastenrath and Greischar, 1997; Mosley-Thompson, 1997). This global compilation indicates that the wastage of mountain glaciers during the last century has raised sea level by between 0.2 and 0.4 mm/year, or roughly 20% of the observed change (Dyurgerov and Meier, 1997; Warrick et al., 1996). Between 2003 and 2009, Himalayan glaciers lost an estimated 192 gigatons of water, and contributed to catastrophic floods of the Indus, Ganges, and Brahmaputra rivers.

Devastating deluge is predicted in areas where glaciers are receding, particularly in the Himalayas (Table 2.4). The number of major glaciers and glacial lakes (many are potentially dangerous) in the Himalayas shows that 20% of Bangladesh will be under water by 2020 if the current rate of glacier melting continues. IPCC (2007) reported that Himalayan glaciers will all melt by 2036, which, however, was later withdrawn as incorrect. According to Bajracharya et al. (2008), the most-likely rate of glacial retreat has been estimated for Bhutan (160 m/year), Nepal (74 m/year), and India (41 m/year). The prediction for 2050 suggests that there will be more frequent glacial lake outburst flood (GLOF), with debris-covered glaciers losing 78–168 m/year and debris-free glacier losing at 20–43 m/year (IPCC, 2013, Table 2.4).

2.2.1.5 Sea Level Change

The GWCC is probably best exemplified by the sea level rise (SLR), which is caused by widespread melting of snow and ice, as well as thermal expansion of sea water as it expands when it warmed above 3.98°C (Kennedy et al., 2010). Water cycles among ocean, atmosphere, and glaciers play important roles in the fluctuation of sea level (Hegerl et al., 2007). In fact, satellite data shows a rise of 3.3 ± 0.4 mm per year from 1993 to 2009 (Fig. 2.11, www.blewbury.co.uk, 2017; Nicholls and Cazenave, 2010). The SLR data from satellite appears to be quite high compared to SLR averaged over the last several thousand years from tide gauge data, which is 195 mm, with 1.46 mm per year between 1870 and 2004 (Church and White, 2006). The measurement shows an average annual rise in sea level of 1.7 ± 0.3 mm per year between 1950 and 2009.

A futuristic estimation by the IPCC (2007) suggested a rise in sea level by 180–590 mm between 2090 and 2099. Using the consistent record of mass balance for the Greenland and Antarctic ice sheets over the past two decades, validated

Table 2.4 Glaciers, Glacial Lakes, and Potentially Dangerous Lakes in South Asia

Countries	Number of Glaciers	Glaciers' Area (sq. km)	Glaciers Ice Reserves (cu. km)	Glacial Lakes (Number)	Glacial Lakes Area (sq. km)	Glacial Lakes Potentially Dangerous
Bhutan	677	1,317	127.31	2,674	106.77	24
India: Himachal Pradesh	2,554	4,161	387.35	156	385.22	16
India: Uttaranchal	1,439	4,060	475.43	127	2.49	0
India: Tista River basin	285	577	64.78	266	20.20	14
Nepal	3,252	5,324	481.32	2,323	75.64	20
Pakistan (Indus river basin)	5,218	15,041	2,738.50	2,420	126.32	52
Total	15,003	33,344	NA	8,799	801.83	203

Source: *Modified after Ives JD, Shreshta RB, Mool PK. 2010. Formation of glacial lakes in the Hindu Kush-Himalayas and GLOF risk assessment. International Center for Integrated Mountain Development (ICIMOD), Kathmandu, Nepal, May 2010. ICIMOD, UNSDR, GFDRR.*

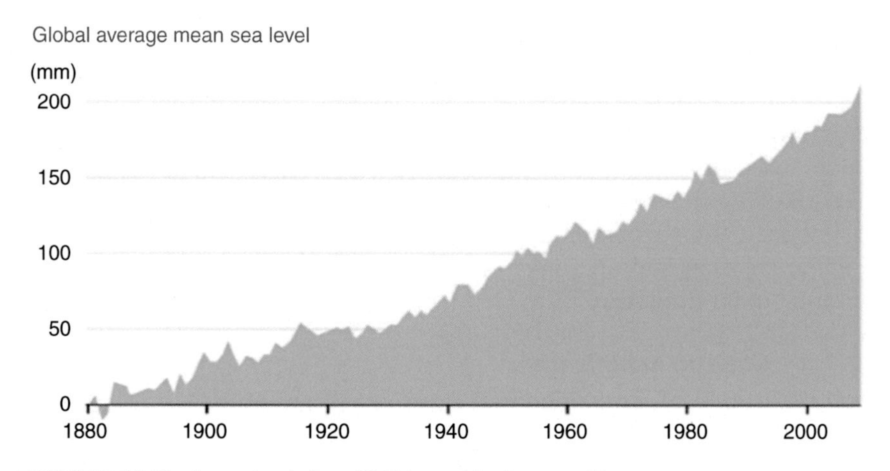

FIGURE 2.11 Rise in sea level since 1880 (www.blewbury.co.uk).

by the comparison of two independent techniques over the past eight years (differencing perimeter loss from net accumulation, and one using a dense time series of time variable gravity), Rignot et al. (2011) predicted a rise of 320 mm in sea level by 2050. The 2014 report of the National Climate Assessment projected a SLR of 300 to 1200 mm by 2100. Relative SLR at specific locations is often 1–2 mm/year greater or less than the global average (>5 mm in the Sundarbans; see later in this chapter). Along the US mid-Atlantic and Gulf Coasts, for example, sea level is rising approximately 3 mm/year (NCA, 2014). This suggests no major acceleration in SLR "with" and "without" anthropogenic contribution (−1.1 to +0.7 mm/year, compared to −1.1 to +0.9 mm/year, respectively; IPCC, 2001).

If the present day sea level and temperature (15°C) are considered as reference points, then during the Eocene period (at 40 Ma), before now the temperature was 19°C and sea level was 75 m above the present-day level. Later, during the Pliocene period (at 3 Ma), the temperature was 17°5C and sea level was 30 m above the present-day level (Fig. 1.2). During the Last Glacial Maxima (LGM at 18,000–20,000 years), the average global temperature was only 9.5°C, while sea level was 120 m below the present-day level. The line at 15°C shows the "most likely" equilibrium response to anomalous global warming (Archer, 2005).

The estimated rise in sea level by 180–590 mm by the year 2100 will submerge several low-lying areas, particularly in the Sundarbans in India and countries such as Bangladesh, Maldives, and Sri Lanka (Harrabin, 2007), including the historical city of Thatta and Badin in Sind, Pakistan (Khan, 2012). The partial submergence of coastal cities (Karachi, Mumbai, Kochi, Male, Colombo, Chennai, Kolkata, and Chittagong) due to rise in sea level could displace several million people (ADB, 2014; Sethi, 2007).

Using geospatial tools and banking on dependable rationale, Dutta (2016) tested the impact of two scenarios of SLR (250 mm and 500 mm) on the coastal regions of populous South Asia. Processing a host of vector data-product from various satellite sources with Arc GIS 10.2 software, potential inundation by integrating digital elevation model (DEM) and Low Elevation Coastal Zones (LECZ) was estimated. The two-prong scenario of what would happen to various parts of South Asia in case the SLRs to 250 mm and to 500 mm is modeled for Andaman and Nicobar Inlands, Bangladesh, east coast of India, west coast of India, Pakistan and Sri Lanka (Dutta, 2016). The results obtained are of concern, but far from scary.

2.2.1.6 *Aquatic Acidification*

About 30%–40% of human-induced emissions of atmospheric CO_2 dissolve into the oceans to form carbonic acid (Caldeira and Wickett, 2003). This change lowers the ocean pH levels turning it acidic (acidification) and threatens the marine ecosystems (Doney et al., 2009). In fact, over the last 250 years, the oceans have absorbed 560 billion tons of CO_2, increasing the acidity of surface waters by 30% (Caldeira and Wickett, 2003; Feely et al., 2009; Orr et al., 2005; Walsh et al., 2014a).

As a result, the world's oceans are becoming more acidic at an alarming rate and could become 150% more acidic by the end of the century, if there is no reduction in emission produced by human activities (Morello, 2010). For example, between 1751 and 1994, the pH of sea surface water is estimated to have decreased from approximately 8.25 to 8.14, representing an increase of almost 30% in H^+ ion concentration in the world's oceans (Hall-Spencer et al., 2008).

At present, the rates of ocean acidification have been about ten times the rate that occurred during the Cretaceous–Paleogene time mass extinction boundary (about 65 million years ago) when surface ocean temperatures rose by 5–6°C (Mora et al., 2013). The effect of acidification on the marine biosphere, although not fully documented, is sure to negatively impact marine shell-forming organisms (e.g., corals and gastropods) and their dependent species.

2.2.2 Impact on Land and Humanity

The increase in GHG in the atmosphere and in water would likely impact the entire society, such as the ecology, environment, economy, agriculture, forestry, ecosystem, water resources, industry, human settlement, and health (Ahmed, 2006; Ahmed and Suphachalasai, 2014). The ever-increasing growth in population accompanied by resource degradation, rising poverty levels, and food insecurity makes South Asia extremely vulnerable to the impacts of such climate change (Sivakumar and Stefanski, 2011; Bhatta et al., 2016). We briefly touch upon the areas of human endeavor, that are expected to be stretched like never before.

2.2.2.1 Food and Agriculture

South Asia needs food security as most of the rural poor depend on agriculture for their livelihood. One of the important requirements of agriculture is availability of water. The irrigation facility and nature of crop influence the yield. Any change in climate, hence, could impact the rainfall that could in turn bring about significant changes in crop yields, production, storage, and distribution. Low-income rural populations that rely on traditional agricultural systems or on marginal lands are particularly vulnerable. Displaying a unique "cycle of consequent damage," the increased concentrations of GHG in the atmosphere would likely to impact agricultural production and practices, which in turn spews out substantial GHG through deforestation, desertification, rice cultivation, enteric fermentation, and fertilizer use.

Crop productivity will be inconsistent and may increase in mid to high latitudes with increase in temperature, but may reduce in the equatorial and tropical regions, if temperature increases beyond 3°C. The IPCC further warns that intense heat waves and decrease in yields of major crops such as wheat, rice, and maize by 2% per decade, against a projected increase in demand by 14%/decade (due to population increase), may enhance the risks from food-borne and water-borne diseases (ADB, 2014).

IPCC (2013) predicted that while crop production in East and Southeast Asia could increase up to 20% by the mid-21st century, the same in Central and South Asia may in fact come down by up to 30% over the same time period. It seems that an increase even by a degree in temperature could cause 20% reduction in yields in South Asia, as rice becomes sterile if exposed to temperatures above 35°C even for an hour during flowering. About 80% of nitrous oxide and almost 100% of terrestrial CO_2 are contributed by agriculture alone. Deforestation, variation in agricultural land cover, burning of vegetation, and consequent albedo effect contribute in increased GHG concentrations, temperature rise, and soot formation (IPCC, 2013).

The change in agricultural land in South Asia over the decades is shown in Table 2.5. In fact, except Bangladesh, all the other seven countries in South Asia have increased agricultural land cover in 2005 compared to that in 1965. Actually, Maldives has doubled the area covered by agriculture, while Bhutan came second. Sharp decline by 4% in land cover used for agriculture in Bangladesh is certainly of concern (IIED, 2011). Climate change and agriculture are very closely related, as agriculture, along with land-use change, enjoys the double distinction of being both a driver and a victim of climate change (i.e., cycle of consequent damage).

Table 2.5A Change in Agricultural Land Cover in South Asia

Countries	1965	2005	Difference
Afghanistan	378,750	379,100	+350
Bangladesh	96,370	93,110	−3,260
Bhutan	3,660	5,620	+1,960
India	1,772,430	1,798,580	+26,150
Maldives	50	100	+50
Nepal	35,310	42,100	+6,790
Pakistan	242,630	270,600	+27,970
Sri Lanka	21,560	25,100	+3,540

All area in km^2.
Source: *Modified after FAO, 2015. Climate Change and Food Systems: Global Assessment for Food Security and Trade. Food and Agricultural Organization of the United Nations, Rome, pp. 1-356.*

Table 2.5B Forest and Agriculture Land Coverage

Country	Land Area (Thousand sq. km)	Forest Area (% of Total Land Area)	Agricultural Land (% of Total Land)
Afghanistan	652	2.1	58.1
Bangladesh	130	11.1	70.3
Bhutan	47	69.1	13.2
India	3287	23.0	60.5
Maldives	0.3	3.0	26.7
Nepal	147	25.4	29.6
Pakistan	796	2.2	34.1
Sri Lanka	65.6	28.8	41.6
Total	5124.9	14.5	33.9

Source: *Modified after Mall and Kumar (2014), UNDP (2013).*

2.2.2.2 Water

More than food or energy or any other resources, water plays an essential role for human survival. Impacts of climate change on water resources will be acute, as it has the potency to dictate growth and migration of population, land-use pattern, economic development, and urbanization. Increased melting of ice from the mountains as well as from polar areas will not only cause rise in sea level and flooding the coastal areas of South Asia, but will also change the salinity balance of the ocean, thus affecting marine life, precipitation pattern, and monsoon. Although ice-melt water will increase the availability of water in the first few decades of temperature rise, flow of water may dry off during the later part of this century due to erratic monsoon and precipitation (IPCC, 2014).

Hence, managing water resources is an extremely important high-end job. Water is subsidized throughout the world, with about 70% of water used for agriculture. About half of 445 rivers in India are too polluted. Again, the extensive southern part of the state of West Bengal is facing acute subsidence due to neo-tectonism in the Bengal Basin, rampant withdrawal of groundwater in an unscientific manner and also the mixing of arsenic with the groundwater (GSI, 2004). The economic, social, health, and environmental costs of such contamination of water are burdening the economy heavily (Letmathe and Biswas, 2015). Again, the Himalayan ecosystems are vulnerable to glacial lake outburst floods and flash floods. Potential increases in evapo-transpiration and rainfall variability are expected to have an adverse impact on the viability of freshwater wetlands, resulting in shrinkage and desiccation.

Availability of fresh water is highly seasonal in South Asia, with about 75% of the annual rainfall occurring during the monsoon months. Melting water from Himalayan glaciers and snowfields currently supplies up to 85% of the dry season flow of the great rivers of the northern Indian plain. This could be reduced to about 30% of its current contribution over the next 50 years, if forecast of climate change and glacial retreat comes true. A reduction in the flow of snow-fed rivers, accompanied by increases in peak flows and sediment yields, would have major impacts on hydropower generation, urban water supply, and agriculture.

2.2.2.3 Pollution

Oil spills, pipeline leaks, submarine accidents, plastics, industrial effluents, earthquakes, mudslides, and volcanic eruptions contribute to the coastal and marine pollution. The most affected coastal systems include wetlands, mangroves, and coral reefs, which provide natural cover to alleviate the impact of coastal disasters. Concentration of heavy metals such as mercury, lead, and cadmium in coastal waters has become a cause of great concern (Chakraborty et al., 2016).

Significant amounts of oil and oil byproducts are released into the environment, mainly due to oil production, transportation, and use, unfavorably affecting marine and coastal environment. Although major oil spills constitute an estimated 2% of the total marine pollution, they cause severe damage to the coastal environment and lead to serious degradation of the shoreline (Nepstad et al., 2017). The major causes of oil spill are causalities such as collisions, or grounding of ships, fires or failure of hull, equipment such as pipelines, flanges, hoses, and incorrect operating procedures.

The toxin produced by the phytoplankton blooms, microalgal blooms, toxic algae, red tides, and harmful algae can contaminate seafood or even kill the fish. Further, these high-biomass producers can cause anoxia at some places to exterminate marine life. The negative effects of harmful algae extend well beyond direct economic losses and human health impairments, as diseases due to the consumption of seafood contaminated by algal toxins are not unusual in the coastal regions of South Asia.

2.2.2.4 Health

The threat of the Earth's average surface temperature exceeding 2°C above preindustrial time is real, so also to the health security of millions of people (Costello et al., 2009). The World Health Organisation estimated that the warming and precipitation trends due to anthropogenic climate change have claimed over 150,000 lives annually during the last 30 years. An upsurge in diarrheal disease, increase in frequency of cardiorespiratory diseases due to higher concentrations of ground-level ozone in urban areas, and the altered spatial distribution of some infectious diseases could prevail in decades to come (EPC, 2016; IPCC, 2014).

In addition to cardiovascular mortality, respiratory illnesses due to heat waves and transmission of infectious diseases will cause health hazards. Moreover, the influence of socioeconomic factors and changes in immunity and drug resistance would play an equally pivotal role. The growing evidence of climate–health relationships suggests increased mortality in many regions of the world and South Asia. Potentially vulnerable regions include the temperate latitudes, which are projected to warm disproportionately (Patz et al., 2005).

Climate change may affect the health in a number of other ways. For example, the increase in the frequency and intensity of heat waves, as well as increase in floods and droughts, will culminate in spreading the vector-borne and diarrheal diseases, primarily in East, South, and Southeast Asia. Human health in these regions will be a direct casualty, as mortality risk for the elderly, chronically sick, very young, and socially isolated persons will increase. Additionally, prolonged exposure to ozone may cause damage to the lungs, and trigger chest pain, skin cancer, cardiovascular complains, and respiratory problems (IPCC, 2014). Increase in temperature will also affect industry and alter demography due to disruption in electricity, water, commerce, transport, and infrastructure. The experience of the 2003 heat wave in Europe shows that not only economically poor countries, but also high-income countries may be adversely affected by climate change. Hence, adaptation to climate change requires well-oiled public health strategies on one hand and effective implementation of standard operating procedures on the other.

2.2.2.5 Infrastructure and Economy

The repercussions of climate change, as indicated above, will be manifested in the fields that directly shape humanity, such as in education, health care, infrastructure, and economic development. For example, it is predicted that by 2050, freshwater availability in the whole of Asia, particularly in large river basins, will reduce. Coastal areas, especially heavily populated megadelta regions, will be at a greater risk due to increased flooding from the sea and rivers (IPCC, 2014). Growth in education, health care, public health initiatives, infrastructure, and economy will help people prepare against any such catastrophe, but would simultaneously require huge financial investment. Hence, a global-scale shift in tourism, passenger, and freight transport is expected.

The predicted rise in sea levels and the associated increase in frequency and intensity of storm surges and flooding would certainly worry coastal population and infrastructure. On the other hand, an increased frequency of drought may lower the water levels considerably, inflicting a negative impact on the inland waterway transport. Climate change will put pressures on natural resources, especially to the environments associated with rapid urbanization, speedy industrialization and economic development. Such growth will make the industries, settlements, and societies located along the coastal and river flood plains vulnerable. In fact, deficit in water is responsible for majority of the problems related to health, transport, and livelihood. This would in turn lead to substantial migration of masses, severely impacting the economy and security of the region (ADB, 2012). In fact, climate change, migration, and security form an interestingly overlapping triangle of anxiety. This interlocking problem is well exemplified in northeast India and Bangladesh (Bhattacharya and Wertz, 2012), where deficit in water and food in one area leads to human migration to the neighboring area, causing in the process security threat in the region.

It is increasingly envisaged in South Asia that climate change response policies (both adaptation and mitigation, A&M) are most effective when they are fully integrated within an overall national development strategy. It is because adequate funding and technological know-how are important in ensuring success of any climate A&M initiatives. For example, six countries in South Asia, such as, Bangladesh, Bhutan, India, the Maldives, Nepal, and Sri Lanka, will see an average economic loss of around 1.8% of their collective annual gross domestic product (GDP) by 2050, falling further to 8.8% by 2100, if the world continues on its current fossil fuel intensive path. The Maldives would be the worst affected, and could lose 12.6% of its GDP annually by 2100, followed by Bangladesh at 9.4%. The losses in Nepal could rise to 9.9% largely because of melting glaciers, while the loss could be 8.7% for India, 6.6% for Bhutan, and 6.5% for Sri Lanka. Put together such losses in South Asia could be higher than in the world. If nothing is done to mitigate the climate change, the global economy could lose 2.6% per year by 2100. For instance, the East Asian economy would lose by 5.3%, the Southeast Asian economy may go down by 6.7%, and the Pacific countries would lose 12.7% of their GDP (ADB, 2016).

2.2.2.6 Coastal Zone

About 38% of the world's population lives within a narrow fringe of coastal land, consisting of only 7.6% of the Earth's total land area. The population here is largely dependent on coastal resources for their livelihood, which are exposed to gradual changes in climate, nutrient loading, habitat fragmentation, and biotic exploitation. Although diverse events can trigger such shifts, recent studies show that a loss of resilience usually paves the way for such switchover (Scheffer et al., 2001). Damage to coastal habitats and wildlife is increasingly becoming more severe in South Asia due to population growth and increased economic and developmental activities. The GHG concentrations in the atmosphere and resultant global warming and climate change (GWCC) could affect the resilience in the coastal ecosystem, as 70% of all the beaches in the world are eroding (Hardy, 2003).

South Asia, with a long coastline of about 12,000 km and extremely high population density along the coasts, has perennially been highly prone to disasters, both, (a) natural, such as atmospheric depression, cyclone, storm surge, tsunami, erosion, sea level rise, and coastal flood, and (b) man-made, such as oil spills, coastal pollution, tourism, fishing, and ballast water exchange. The anthropogenic intervention to the natural ecosystem will impact coastal vulnerability and demographic dynamics in South Asia. Development in this part of the world has been undertaken too close to the sea (haphazardly and menacingly, at times in violation of legal provisions), particularly in several megacities. For example, Karachi, Mumbai, Cochin, Colombo, Male, Chennai, Kolkata, Visakhapatnam, and Chittagong have proliferated rapidly over the last few decades, compared to several comparatively smaller yet crowded cities.

The Ganges and Brahmaputra contribute a major share of suspended sediments to the Bay of Bengal, so also the river Indus to the Arabian Sea. Any attempt either to arrest erosion or prevent deposition requires a thorough understanding of the factors and processes involved in the coastal geomorphologic system and input of sediment transported by rivers to the estuaries. The erosion–accretion dynamics of two major estuarine regions of South Asia—the Sundarbans at the mouth of the river Ganges and Brahmaputra along the east coast and the Gulf of Khambhat at the mouth of the river Narmada along the west coast of India are specially examined here.

2.2.2.7 Ganga-Brahmaputra Delta (Sundarbans)

The Sundarbans, roughly spanning between longitude 88°E to 90°E and latitude 21°30′N to 22°30′N, spreads over the two neighboring countries of India and Bangladesh. This area is affected by the monsoon and tidal current, and is very often battered by cyclonic wind, flood, storm surge, and submergence. Data on 16 forcing parameters comprising atmospheric, oceanographic, and terrestrial components were acquired continuously over a period of little more than three decades (1979–2011) by Chatterjee et al. (2015). These data were obtained from fourteen southernmost sea-facing islands of the Sundarbans to study the coastal vulnerability in terms of variation in shoreline pattern, erosion–accretion dynamics, and change in coverland-use pattern (Fig. 2.12).

The digitized temporal shoreline shift of six western islands (Ghoramara, Sagar, Jambudwip, Mousuni, Namkhana, and Lothian) during more than three decades indicates maximum erosion on the northwestern and western banks of these islands, compared to southern and eastern banks (Table 2.6). Of the fourteen islands, five had loss of minor area, varying between 1.3% and 8.5% between the years 1979 and 2011 (least vulnerable islands). Another three islands recorded moderate vulnerability and loss of land to the tune of 12.6% to 21.1%, while the remaining four islands maintain evidences of major loss of land area by about 32% to 40% during the same period (highly vulnerable). While one island (Ghoramara) was completely submerged by the mid-90s, the Nayachar island showed a continuous accretion since 1979 and nearly doubled its land area. The location of Nayachar island on the healthier

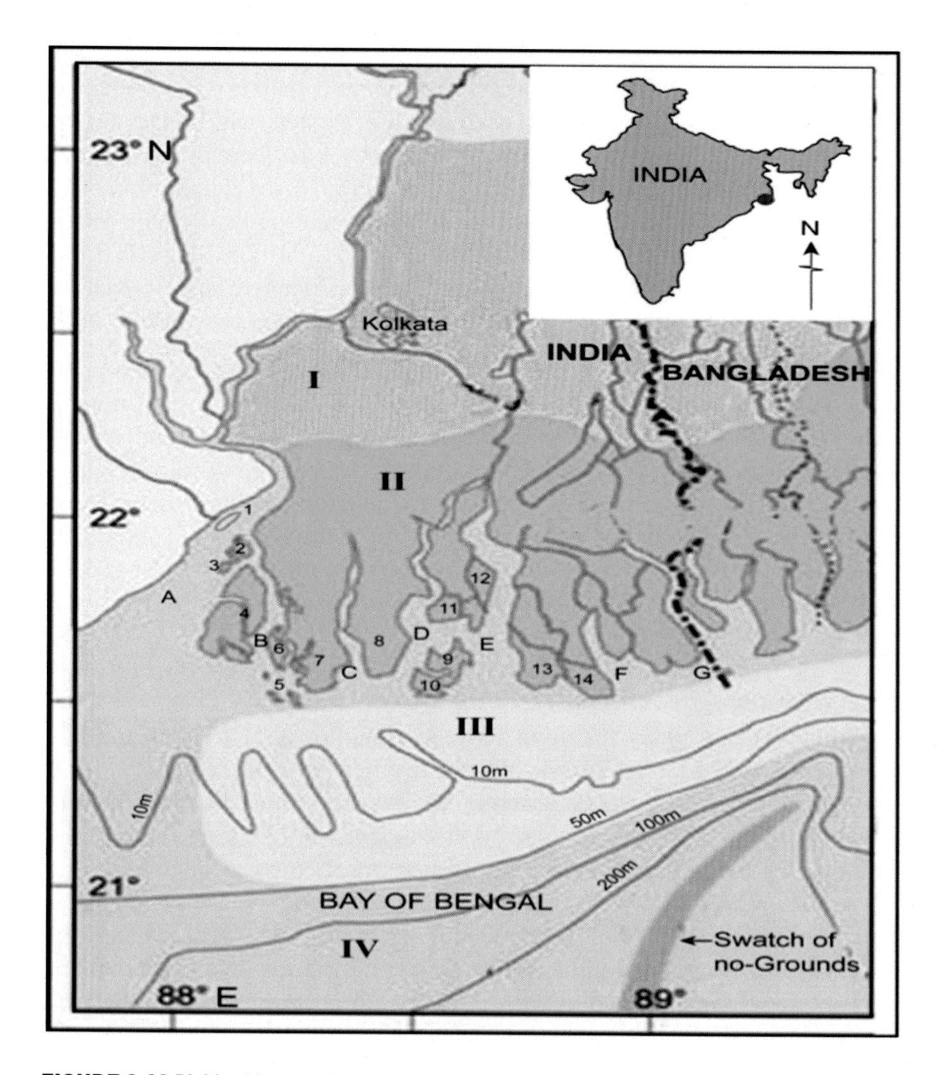

FIGURE 2.12 Digitized image of a part of the Sundarbans Deltaic region.
Figure processed from LANDSAT-3 MSS data. The *black north-south line of dash and dot* represent political boundary between India and Bangladesh. I = Mature delta, II = Tidallyactive delta, III = Subaqueous delta, IV = Open sea. The rivers are designated as: A = Hooghly, B = Muriganga, C = Saptamukhi, D = Thakuran, E = Matla, F = Bhangaduani, G = Goasaba. The islands are: 1 = Nayachar (*only accreting island in Sundarbans*), 2 = Ghoramara, 3 = Lohachar (*presently submerged*), 4 = Sagar Island (*pilgrimage centre*), 5 = Jambudwip, 6 = Mousuni, 7 = Namkhana (*tourism center*), 8 = Lothian, 9 = Chulkati, 10 = Bulchery, 11 = Surendranagar, 12 = Dhanchi, 13 = Dalhousie, 14 = Bhangaduani. The islands 13, 14, and their northern areas encompass national park and sanctuaries. The areas coming under islands 8–12 and their northern extensions are covered by deep forest. *Thin lines* are bathymetry contours. "Swatch of no-ground" is a sharp, deep nearly north-south trending narrow gorge. Haldia Port is located on the western bank of river Hooghly. *(After Chatterjee, N., Mukhopadhyay, R., Mitra, D., 2015. Decadal changes in shoreline pattern in Sundarbans. J. Coast. Sci. 2 (2), 54–64).*

Table 2.6 Gain and Loss of Land Area in Sundarbans

Island	1979	1989	2001	2011	Loss %	Land Loss	Erosion Direction	Accretion Direction
Extreme loss								
Lohachar	3.42	1.47	0	0	100	3.42	N, S	SE
Major area loss								
Ghoramara	7.40	6.38	5.21	4.41	40.40	2.99	NW, SW, NE, SE	–
Bhangaduani	40.18	37.79	30.74	25.00	37.77	15.18	S	–
Bulchery	32.20	26.30	23.12	21.20	34.16	11.00	S	
Jambudwip	7.34	7.14	5.85	4.95	32.56	2.39	S, W, N	E
Moderate area loss								
Dalhousie	76.51	70.70	66.58	60.32	21.16	16.19	S	
Chulkati	44.00	42.00	41.76	38.30	12.95	5.70	S	
Mousuni	30.81	30.00	28.62	26.90	12.60	3.91	NW, N, W	
Minor area loss								
Dhanchi	36.68	35.47	34.44	33.56	08.50	3.12	S	NNE
Surendranagar	36.02	35.57	34.41	33.36	07.38	2.66	S, SE, SW	W
Namkhana	151.63	150.2	147.30	145	04.37	6.63	NW, S, W	E
Sagar	242.07	237.82	238.92	234.72	03.03	7.35	S, SW	SE, N
Lothian	35.02	33.81	33.05	34.56	01.31	0.46	W	E, S, N
Area gained								
Nayachar	32.33	42.2	53.23	64.14	Nearly double	31.81	SE	SW, W

All figures in km². Major loss >30% area, moderate loss = 10%–30%, minor loss = <10%. Total loss of area in six Western islands is 23.73 km² (5.01% of the original land; average loss per island 15.71%) and in six Eastern islands is 53.85 km² (20.28% of the original land; average loss per island 20.32%.

Source: *From Chatterjee, N., Mukhopadhyay, R., Mitra, D., 2015. Decadal changes in shoreline pattern in Sundarbans. J. Coast. Sci. 2 (2), 54–64.*

western channel of river Hooghly, subsurface tilt of this area toward the east, and less aggressive tidal impact might have played important role for its growth amidst ruins (Chatterjee et al., 2015; Ghosh et al., 2003). The satellite images derived from IRS 1D LISS III showed emergence of another island called Balari-Bar in the river Hooghly, which clearly indicates the depositional regime in the western channel of the Hooghly river.

The erosion–accretion dynamics suggests that the total area of Sundarbans has dwindled by about 57% within a period of the last 200 years (Ganguly et al., 2006). Another set of data encompassing a period between 1840 and 1980 (span of 140 years), however, reported accretion of shoreline by about 2–5 km^2/year in the eastern Sundarbans (areas now under Bangladesh), while the western part (Indian Sundarbans) recorded erosion by 1.9 km^2/year (Allison, 1998; Jayappa et al., 2006). Further, it is also suggested that the western part of the Indian Sundarbans has lost about 250 km^2 of land during the last 80 years (Bhattacharya, 2008; CSE, 2012).

Incidentally, there appears to be an inverse relation between size of the island and percentage of land loss. For example, the islands with originally more than 150 km^2 (e.g., Sagar and Namkhana) show only 3% to 4.5% of land loss. In contrast, smaller islands with original area less than 25 km^2 (e.g., Lohachar, Jambudwip, and Ghoramara) faced severe erosion and lost land to the tune of 32% to 100% (Table 2.6). The rest of the islands with original size between 25 and 150 km^2 record a moderate land loss, varying between 7% and 38%. The results appear to suggest that larger islands could withstand the impact of high tide surge and circulation of long-shore currents. Increased tide height and cyclonic activities in the Sundarbans over the last few decades have damaged the topsoil, killed cattle, and rendered people homeless to become environmental refugees (Gopal and Chauhan, 2006; Kotal et al., 2009).

The difference in SST and salinity is causing retardation in seed germination and an impediment to the growth of mangroves (mangroves are only salt tolerant and not salt lovers), resulting in increased erosion of land in the eastern islands (Chowdhury, 1987; Mirza, 1998; Mukhopadhyay et al., 2006; Sarkar et al., 2002). In addition, cyclone and storms with speed ranging between 80 and 120 kmph or beyond are important atmospheric parameters to cause rapid and large-scale erosion (Michels et al., 1998). In fact, between 1961 and 2000, as many as 31 severe cyclones occurred in the Sundarbans, of which at least in 5 cases the speed reached over 200 km/hour. It seems that the occurrence of cyclones in the Sundarbans has increased by about 26% during the last 120 years (CSE, 2012). Moreover, the Sundarbans is a tide-dominated delta system, where tidal heights reach up to 5.3 m, and inundate further inland.

A large part of the Sundarbans also falls under the highly active zone IV in the seismic map (GSI, 2004), where episodic neotectonic movements occur, with latest ones occurring about 900 and 200 years ago, which resulted in an easterly tilt of the Bengal Basin (Curray et al., 1982; Mirza, 1998). The local SLR in Sagar Island in the Sundarbans and nearby Diamond Harbour has been reported to be 5.22 mm/year and 3.14 mm/year, respectively (Unnikrishnan and Shankar, 2007). Both these values are much higher than the Indian national average of rise in sea level at 1.88 mm/year. Such atypical rise in sea level in the Sundarbans may have probably caused by subsidence, coupled with tilting and isostatic sinking of the delta, large-scale wind action, and low elevation of the coast (Bhattacharya, 2008; Curray et al., 1982; GSI, 2004; Ganguly et al., 2006; Kotal et al., 2009).

The tidal incursion upstream and cyclonic inundation of coastal land by seawater are causing widespread conversion of several mangrove areas into aquaculture in the Sundarbans. An increasing population in the Sundarbans (from 0.296 million in 1872 to close to 4.5 millions in 2011) and a high population density of 904/km^2 (Census of India, 2011) are adding huge pressure on mangroves, fisheries, agriculture, and land-use pattern (Datta, 2010; Roy, 2010). The impact of anthropogenic activities appears to have played an important role in the western part, where a major pilgrimage center (Sagar) and two tourism centers (at Frazerganj and Bakkhali) are located.

The demographic pattern in the Sundarbans has undergone considerable modification over the last few decades because of the variable concentration of heavy metals in organisms and the presence of high concentrations of arsenic in the ground waters (Milliman and Meade, 1983; Mukhopadhyay et al., 2006; Sarkar et al., 2002). Clayey sediments intercalated with sandy aquifers at depths between 20 and 80 m are reported as a major source of arsenic in groundwater (Bhattacharya et al., 1997). It is considered that the rise in sea level during Holocene and the development of reducing conditions at organic-rich swampy lands are directly linked to epicenters of arsenic distribution (Shamsudduha and Uddin, 2007). As this arsenic-contaminated water is being used for irrigating agricultural farms, fish-culture ponds, and also for use as potable water for the multitudinous residents, the threat to human health has become stupendous. Also the balance of water ecosystem in the Sundarbans is disturbed by eutrophication, that is, excessive fertilization (nutrient enrichment), which, in turn, leads to increase in phytoplankton production, creating in the process an anoxic condition and damaging the mangrove vegetation (Chaudhuri and Chaudhury, 1994; Pernetta, 1993; Ray, 2008).

In addition, a large number of people of the Sundarbans exploit mangrove vegetation for food, fodder, fuel, and medicine. To integrate climate change policies to the sustainable development strategies in general, and poverty alleviation measures in particular, identifying the causes for land-use change in the Sundarbans is of immense importance. Such campaign will relate the economy of the area to the consequent pattern of migration of people (Anand and Sen, 1994; Datta, 2010; Roy, 2010). Although about 78% of the economy of the Sundarbans depends on agriculture, the availability of per capita land for agriculture has been constantly reducing (CSE, 2012). In fact, the conversion of paddy plus mangrove lands to aquaculture and further transfer from aquaculture to settlement zones in the Sundarbans has been rampant—maximum in Sandeshkhali (104 km^2 and 17 km^2, respectively) and minimum in Kakdwip (8 km^2 and 3 km^2; Chopra et al., 2006).

2.2.2.8 Narmada–Tapti Estuary (Gulf Of Khambhat)

The Gulf of Khambhat, located on the western coast of India, is a region that holds great interest due to its unique geomorpholgy, rich biodiversity, and thriving industrial activity in Surat and Mumbai (Fig. 2.13). Major rivers such as the Narmada, Tapi, Sabarmati, Mahi, and Shetrunji and many other minor rivers debouch into the Gulf, with its flow bringing large loads of sediments.

Because of its distinct inverted funnel-shaped structure, the Gulf of Khambhat has been dominated by some of the highest energy tides in South Asia (max tidal height 12.5 m). Additionally, being a low elevated coastal zone (<10 m), the Gulf and the associated regions are vulnerable to rising seas. Mangrove degradation here is one of the worst in Gujarat, and salinity intrusion has deleteriously impacted the agricultural productivity even in hinterland areas. The rising trend of cyclonic activity in the Arabian Sea and rising SSTs are a cause of grave concern for populations in this region.

The Gulf covers an area of about 3120 km^2 comprised mainly of mudflats and a volume of about 62,400 million m^3 of water (ICMAM, 2002). Rainfall varies annually from 600 mm on the western side to 800 mm on the east. Humidity ranges between 65% and 84%. Setting up of industries and building of ports

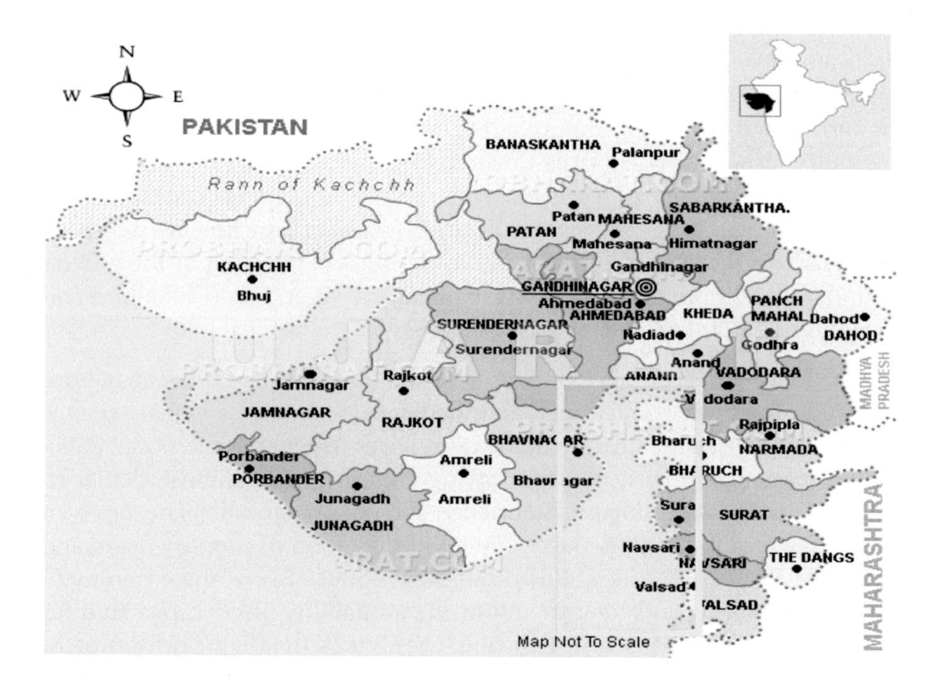

FIGURE 2.13 **Index map of coastal areas in the Gulf of Khambhat in the Arabian Sea.** *Yellow box* is the study area *(Modified after Davis, G.T., 2012. Vulnerability Assessment of the Gulf of Khambhat. Dissertation Thesis, National Institute of Oceanography, Goa, India. 78).*

and shipyards have also altered the nature of the coast in many areas. Increasing tidal activity is also a matter of concern. All these forcing factors make this coast and its ecosystem vulnerable to oceanic and anthropogenic activities and, hence, were studied continuously over a period of four decades (Davis, 2012).

Lack of rainfall, intrusion of saline water during high tide incursion, and rampant spreading of the salt industry along the coast has left vast areas of land barren and has caused salinity ingression in the groundwater aquifers that once used to provide drinking water. Agricultural produce has diminished in many regions with people having to shift from traditional farming crops such as bajra and groundnut to salt-tolerant crops such as cashew and cotton (Davis, 2012), thus even affecting the source of food for people in these regions. Availability of drinking water is one of the biggest challenges faced by them. The shortage of drinking water had deleteriously impacted the health of the people, as six villages showed fluorosis problems caused by salinity ingression. Among these six villages, five were located along the coast. It tends to affect the agriculture, economy, and health of the people staying in the area (Davis, 2012).

Partially treated and mostly untreated effluents from industries, unrestricted cutting of mangrove by locals for fodder and fuelwood, and reclamation of mangrove land for salt pans and other industries are the major concerns in the Gulf of Khambhat (Hirway and Goswami, 2007). As a corrective measure, efforts are being made by the local government in collaboration with NGOs and neighboring communities to replant mangroves along the coast. In fact, there has been an overall increase in mangrove coverage by about 31.3% between 1998 and 2011. Salinity ingression in the Gulf is a serious problem with districts like Bhavnagar, Ahmadabad, Bharuch, Surat, and Valsad facing the worst.

Shoreline change analysis revealed that most of the areas of the Gulf of Khambhat are relatively stable and are not experiencing much change (except probably erosion of 6.6 kms of the Valsad coast over the past 45 years). The greatest change in terms of erosion and accretion has been encountered mainly at the mouths of the rivers, which are areas of ceaseless dynamic change. The accretion seen at the mouth of the Narmada, including emergence of Aliya Bet island, is because of huge sediment load that the river brings in.

2.3 A QUICK APPRAISAL

As plethora of data and information are made available by various government and nongovernment agencies and research institutions with regard to causes and impact of climate change, it would be ideal here to take a quick stock of the situation. While global warming refers to an increase in average temperatures, the climate change refers to changes in seasonal temperature, precipitation, wind pattern, and humidity for a given area. Climate change does not necessarily mean warming only. While most climate change generally occurs slowly over time, there is evidence that episodes of rapid cooling have also occurred in the past. Scientists have found evidence that a fall in temperature occurred twice within the past 12,700 years. The natural ocean conveyor belt (NOCB) is found to help regulate the planet's climate.

While some scientists continue to believe that global warming is due to changes in intensity of solar radiation and sunspot activities, the majority, however, agree that the recent changes in temperature are unlikely to be wholly natural in origin. Instead, they believe that the warming being experienced today is due to the rising concentrations of heat-trapping GHG that form a blanket around the Earth. These gases are put into the atmosphere primarily by human activities—particularly the burning of fossil fuels. As a result of global warming and climate change, some regions—such as Siberia—will likely become warmer and more habitable, while in some other areas, spring may arrive earlier and winter sets in later.

In fact, ocean produces approximately half of the oxygen available on the Earth. However, the combined effects of nutrient loading and climate change are creating "dead zones" (oxygen depleted zones) in the open ocean and in coastal waters. The continuing increase in number and size of such dead zones is alarming. The International Oceanographic Commission through its project Global Ocean Oxygen Network (GO2NE) is studying to keep oxygen levels in check, protect vulnerable marine life, and improve low oxygen levels in water in an attempt to counter the spread of dead zones in the ocean. Although the rate of deoxygenation appears to be lower in the seas around South Asia than in the Atlantic and Pacific oceans, Indian Ocean, especially the Bay of Bengal, is extremely vulnerable to even minor oxygen declines. The western shelf of India also houses the world's largest shallow water low-oxygen zone more such zones could develop in the Indian Ocean due to high population pressure. If this happens, it will have far-reaching ramifications on fisheries economy, human health, and other socioeconomic areas. To counter such oxygen depletion, it is suggested to adopt better sanitation system to protect human health and pollution, to curb fossil fuel burning to cut GHG emission, and to create marine-protected areas or no-catch zones in locations that are used by animals to escape low oxygen. Hence, while mitigating climate change requires a global effort, few local actions can reduce nutrient-driven oxygen decline.

Wealthier and developed industrial countries contribute the most to global warming because they use most of the world's fossil fuels. Europe, Japan, and North America—even with only 15% of the world's population—are estimated to account for two-thirds of the CO_2 now in the atmosphere. With less than five percent of the world population, the United States is the single-largest source of carbon emission from fossil fuels—emitting 24% of the world's total. The US automobiles (more than 128 million or one-quarter of the world's cars) emit roughly as much carbon as the entire Japanese economy, the world's fourth-largest carbon emitter in 2000.

Concentration of PFC, SF_6, HFS23, and C_2F_6 in the atmosphere of South Asia is very low, so also that of SO_2, CH_4, CO_2, and N_2O, compared to that in United States and Europe. Again, aerosols are found to be more abundant in the atmosphere of the northern hemisphere than in the southern latitudes. Its concentration in Europe and America, however, shows a recent downward trend, but the same is increasing in South and East Asia, although at a much lower rate than the world average.

However, the effects of all of these GHG and aerosols on the Earth's climate depend in part on how long they remain in the atmosphere (Table 2.1). About one-third of the CO_2 emitted in any given year remains in the atmosphere for 100 years; however, its impact remains for tens of thousands of years. Methane lasts for approximately a decade before it is removed through chemical reactions. Particulate matters (PM), on the other hand, remain in the atmosphere for only a few days to several weeks. Therefore, some quick mitigating actions to reduce PM contribution can show results nearly immediately, while that for mainstream GHG may take decades (Shindell et al., 2012). Observational evidence shows that heat-trapping gases (CO_2, CH_4, SO_2, N_2O, etc) other than water vapor together account for between 25% and 33% of the total greenhouse effect (Lacis et al., 2010; Schmidt et al., 2010).

During the last three decades, direct observations indicate that the Sun's energy output has decreased, albeit slightly. The two major volcanic eruptions of the past 30 years also have had only short-term cooling effects on climate, lasting for two to three years. Yet, 15 of the 16 warmest years on record have occurred since the beginning of the 21st century, indicating that present change in climate is unusually fast compared to past changes (USEPA, 2016; IPCC, 2014). The human influence on the climate system—including GHG emissions, atmospheric particulates, and land-use and land-cover change—could help to explain acceleration in recent times (USGCRP, 2014; US-EPA, 2016).

In summary, we find a general consensus within the scientific community that climate is changing. Although primarily contributed by natural causes, GHG emissions from human activities are making the climate change faster, visible, and painful.

CHAPTER 3

Political Ecology

The earlier chapters have shown that the climate was, is, and will continue to change. The question is to what extent humans are responsible for such a change? The level of uncertainty in IPCC models questions the very speculations made by this worlds' top scientific body. Nonetheless, even a conservative global warming and climate change (GWCC) in South Asia could hinders the economic growth, agricultural development, and human livelihood. Hence, it is thought to be necessary to vie for soft adaptation and risk management, and to opt for some low-carbon growth options. Such measures may open up new job opportunities as well as innovations in technology. These initiatives would enhance international cooperation and collaboration, and bring about, in the process, a change in the mindset—live for everybody's need, not for one's own greed.

Another issue that makes global negotiations challenging, with specific reference to South Asia, is the aspect of availability of funds for adaptation and mitigation. A conservative estimate suggests the requirement of US$100 billion annually for adaptation measures. But "Who should pay this and to whom"—has been the major bone of contention, particularly in the absence of undisputed data on the degree of climate change impact a country is subjected to. The emergence of various smaller power centers and groups of economic and industrial nations, such as, G-5, G-8, G-15, and BRICS, brings about "conflict of interest" in working out a robust strategy to arrest climate change.

We all know that the GWCC is caused by both natural forcing parameters and man-made (anthropogenic) activities. Yet, highlighting only man-made activities as responsible for climate change by the United Nations Framework Convention on Climate Change (UNFCCC) may be considered biased by few. Incidentally, some of the assumptions that man-made activities are responsible for warming of the globe are true, and some are not. Gradually, the debate is turning into a case between silent majority and vocal few. The commercial interest, which indisputably influences the global politics, is considerably

Climate Change. http://dx.doi.org/10.1016/B978-0-12-812164-1.00003-7

involved in the issue and is alleged to slowly outweighing other considerations. It is also apprehended that narrow interest groups may aggressively push for public policies that are asymmetric in nature (Haller and Gerrie, 2011). It is apprehended that when science enters the arena of politics, it loses its sheen of integrity.

Again, in addition to acquiring and assessing strong and irrefutable scientific data before framing any law and implementing any measure, one must also give importance to culture, belief, customs, and socioeconomic conditions, wherein such measures are to be implemented. The entire local community of people of the concerned area needs to participate willingly in such ventures. People may be encouraged to take extra interest in understanding, spreading, and leading the palliative and adaptive measures. Although industries and some governments witnessed economic disaster in implementing Kyoto Protocol conditions, scientists (even some notable economists) believe that the world has missed the best opportunity by allowing it to lapse.

Taking quite a different view, a considerable number of scientists argue that the Earth is on a cyclic mode and the present rise in temperature is happening after a prolonged thousands of years of cooling. This debate has generated two schools of thoughts (IPCC vs NIPCC) among the three worlds (developed, developing, and low developed countries) over the pattern, estimate, and rate of warming of the globe through human interference. In this chapter, we briefly discuss various thoughts and hypotheses with regard to carbon politics, and what those mean to South Asia.

3.1 CARBON POLITICS: TWO SCHOOLS

3.1.1 First School: Globe is Warming

This school is being led by the Intergovernmental Panel on Climate Change (IPCC) through several of its assessment reports (1990, 1996, 2001, 2007, and 2014). These reports find beyond reasonable doubt that each of the last three decades have been successively warmer than any preceding decades since 1850. The IPCC claims that the average increase in the temperature of the Earth's surface has been 0.85 °C. Globally, sea levels have risen faster than at any other time during the previous two millennia—and the effects are felt in South Asia. Driven by shift in surface wind pattern, thermal expansion of ocean water, and addition of melting ice, ocean currents are altering sea level that varies from place to place. Gravitational field of the Earth also cause regional fluctuations in sea level. Additional variations are caused by sediment and tectonics. Changing patterns of rainfall or melting of snow and ice are altering the freshwater systems, affecting the quantity and quality of water available in many regions of the globe.

The IPCC has asserted that human-induced activities are the dominant causes behind global warming since the mid-20th century. Given the interdependence among the countries in today's world, the impacts of climate change on resources or commodities in one place will have far-reaching effects on prices, supply chains, trade, investment, and political relations in other places. IPCC (2014) further observes a decrease in cold days and nights in South Asia, with an increase in heat wave frequency since the 1950s. Rainfall in South Asia has been extremely erratic and bears no definite pattern. Instead of a robust climate, South Asia is plagued during the last few years by incessant rain, cloud burst, and extended drought—a condition that never existed before. While most areas of the Asian region lack sufficient observational records to draw conclusions about trends in rainfall, flood, and drought, these calamities take a heavy toll of South Asian economy in terms of life, property, infrastructure, and health (diarrheal outbreaks and cholera).

It seems the global surface temperature has been increasing steadily by about 0.13°C (0.10–0.16°C) per decade. The temperature increase is greater at higher northern latitudes, and more on land than in ocean. Yet, ocean to a depth of even 3000 m has been warmed-up, as it is absorbing about 60%–80% of increased heat in the atmosphere. Both hemispheres recorded decrease in glaciers and snows in the mountain, varying from 7 to 15%. Consequently, the global sea level has increased by about 3.1 mm/year (range 2.4–3.8) between 1993 and 2003 (Fig. 2.11). Using the increase in atmospheric CO_2 levels from 370 ppm to 400 ppm during the past decade, the IPCC suggested warming of the world by at least 0.2°C following changes in solar irradiance, volcanic eruptions at various places, the Pacific Decadal Oscillation, the North Atlantic Oscillation, and the Arctic Oscillation.

3.1.2 Second School: Globe Is Not Warming

This school is led by the Non-Governmental International Panel on Climate Change (NIPCC, 2010), which highlights the weaknesses of the IPCC reports. NIPCC finds the probability by IPCC that increased temperature will increase sea level mainly due to melting of the ice-caps untenable, as in the past one century, the sea level has risen by only 20 cm, and there is no evidence for its acceleration in the future. Moreover, in contrast to the claim of IPCC, there has been no increase in frequency of hurricane activity in the last 40 years, and modest variation, if any, in sea level and frequency of hurricane does not warrant drastic action.

In fact, a decrease in both the intensity and number of hurricanes is recorded after 1940 in the Indian Ocean (Fig. 2.7). In fact, between 1991 and 1995, relatively few hurricanes occurred, in contrast to the IPCC report (1996), suggesting that some regions may experience increased cyclonic activity while others will see fewer. Again, because wind speed at various altitudes seems to play a larger role than the temperature, it is possible that any regional changes in cyclonic activity will continue to occur against a backdrop of large yearly natural variations (Idso and Singer, 2010).

The NIPCC also claims that there is no evidence to suggest that global warming is dangerous to the Earth, and points out that warming, even if true, is beneficial to biodiversity, human health, and environment. Quoting an interesting study of Jaramillo et al. (2010) on pollen grains and other biological indicators of plant life embedded in rocks formed during abrupt warming period during Paleocene-Eocene Thermal Maximum (PETM) at about 56 Ma, NIPCC observes that contrary to expectations, the forests bloomed with diversity when CO_2 levels had doubled and the world was warmer by 3–5 °C. New species of plants evolved quicker as others became extinct, as they acquire genetic variability to cope with high temperature and high levels of CO_2 (Jaramillo et al., 2010).

Further, NIPCC emphasizes that anthropogenic carbon emission has a miniscule influence on global warming, and hence, industries of the developing economies like India should not be disproportionately burdened to cut carbon emission (Liberty Institute, 2002). The inconsistencies in the claims and arguments against it have forced the alarmist group to change the slogan to climate change from global warming. Barnosky (2011) and the IUCN Report (2012) also hinted at the policymakers to avoid acting in haste, as the proposed costly actions to reduce GHG may not be justified by the best available evidences.

In fact, the NIPCC argues that there is no scientific consensus that global warming is a problem, or that humans are its cause. Even if current predictions of global warming are correct, delaying drastic government actions by up to 25 years will not make much difference to global temperature 100 years from now. In contrast, postponing action until the world acquires adequate, repetitive, and reliable evidences that human activity is causing global warming, and developing better technology to mitigate the problem, will make both the environment and economics a good sense (Idso and Singer, 2010).

Ashworth et al. (2011) and van Andel (2011) argued that the perception that CO_2 is the main cause behind global warming may not be completely true. In fact, all GHG, including the most infamous water vapor and CO_2, cool the planet. Hence, any cap on taxing carbon emission will have a negative impact on the economies of the world. Zeigler (2013) concluded that increase in CO_2 and temperature can be readily explained by natural processes. For example, apart from orbital forcing and the distribution of continents and oceans, fluctuations in solar activity and the galactic cosmic ray (GCR) flux did control climate changes during the geological past, and probably is still regulating. According to NIPCC, there is overwhelming evidence that temperature forces the carbon cycle and not vice-a-versa, as mistakenly postulated by IPCC (Idso and Singer, 2010; Zeigler, 2013).

More evidences are now emerging to suggest that the oceans are absorbing more heat than what was estimated so far. Temperatures at depths greater than 700 m appear to have increased more than ever before, whereas it was believed till now that heat is absorbed only up to the top 300–400 m of water column. IPCC simulations failed to predict this effect.

The other possibility is that IPCC in its simulations might have underestimated the climate fluctuations owing to natural causes (Hansen et al., 2013; Nikolov and Zeller, 2011; Solomon et al., 2010). The fact that the world has not warmed at all means that all the other climatic factors have had a net effect of producing 0.2°C of cooling (*david.whitehouse@thegwpf.org*). Sea level measurements for Tuvalu, a tiny archipelago in the western Pacific Ocean between Hawaii and Australia (and 10 other stations) between 1992 and 2006 show that for about the past eight years the sea level seems to be virtually constant (IPCC, 2014). Another study reports that the east Antarctic ice sheet north of 81.60°S, in fact, increased in mass by 45 ± 7 billion metric tons per year from 1992 to 2003. This was enough to slow down the sea level rise by 0.12 ± 0.002 mm/year (Davis et al., 2005). It is shown that 72% of the Antarctic ice sheet is gaining that could lower global sea levels by 0.08 mm/year (Wingham et al., 2006).

3.2 CARBON POLITICS: THREE WORLDS

The climate negotiation has been going through a rough patch for the last three decades, with several new dimensions emerging during every conference. In fact, global negotiations are required because climate change is a global challenge and requires a global solution. The GHG emissions would have the same impact on the atmosphere whether they originate in New Delhi, Canberra, Washington, London, Paris, Moscow, or Beijing. Consequently, initiatives taken by one country to reduce emissions will contribute negligibly to slow global warming unless other countries act as well. A brief history of negotiations at highest level is tabulated in Table 3.1.

The politics on carbon emission and carbon credit divided the world into three subtle worlds—first world (developed), second world (developing), and the third world (least developing countries). It would be rather appropriate to reflect briefly on the type of negotiations followed globally among these three worlds, for the reason that negotiations are likely to influence the government policies. The saga started with the formation of the United Nations Framework Convention on the Climate Change (UNFCCC) during the Rio de Janeiro Earth Summit in 1992. It came into force on March 21, 1994 and has been ratified by 196 countries, which constitute the "Parties" to the Convention—its stakeholders. The member-nations of the three worlds took strikingly different positions with regard to various aspects of climate negotiations (Table 3.2).

The "Kyoto Protocol" was adopted in 1997 for a 15-year-term and was considered as the basis to mitigate future GHG emissions. The protocol suggested to: (a) set up a "global climate fund" to help the poor nations, (b) create a mechanism to share the clean technologies, (c) protect tropical forests, and (d) help the poor adapt impacts ranging from storms to rising sea levels.

The Kyoto Protocol was to get renewed in 2012, but by 2009 during the Copenhagen summit, it was apparent that this protocol had fallen short of its objectives, and there was a need to undertake a new negotiation. The summits of the UNFCCC in Cancun-2010 and Durban-2011 tried their best to reach a consensus before June 30, 2012 when the Kyoto Protocol was to become lapse, but both failed. The Cancun-2010 UNFCCC meeting may have failed but not before offering a mixed bag of solutions for all the three worlds. While the meet agreed to India's proposal for greater transparency in mitigation planning and actions, many other countries opposed India's initiative to maintain the Kyoto Protocol. Under the Kyoto Protocol, the first-world (developed, rich nations) had to reduce emissions while developing poor nations had to take nonbinding voluntary mitigation actions. In fact, the differences were so substantial that the ad-hoc working group had to ultimately release two separate drafts on the mitigation and transparency, one each for the developed and developing worlds (Chauhan, 2010). The suggestion of Cancun 2010 for the developed nations to start paying by 2020 over US$100 billion every year in climate aid to poor countries to help them take appropriate measures to hold the increase in global average temperature below 2°C above preindustrial levels further distanced the nations.

While countries of BRICS (Brazil, Russia, India, China, and South Africa) maintained the Kyoto Protocol as nonnegotiable, a new demand by the Island nations and few of the least developed countries, led by Bangladesh, and supported by EU and USA, was put forward to bind "all countries" including BRICS to cut emission. This new demand had threatened to divide even the developing world. The three developed countries—Japan, Canada, and Russia, however, insisted for a single binding agreement for all the nations to cut GHG emissions so as to hold the increase in global average temperature below 2°C above preindustrial levels.

Table 3.1 Tabulated History of International Negotiation

COP	Discussion and Outcome		Discussion and Outcome
COP-22	Marrakesh, Morocco, November 7–18, 2016, took note of progress made after the Paris Agreement. Reaffirming cooperation, the COP-22 finalized a Partnership for Global Climate Action.		
COP-21 Paris, France, November 30–December 11, 2015	Must agree to limit global warming below 2°C. Must frame a legally binding agreement toward resilient, low-carbon societies and economies. The agreement is to come in force from 2020. Must mobilize GCF of $100 billion per year from 2020 to help developing countries to combat climate change. France contributed US$1 billion	COP-11 Montreal, Canada, November 28–December 10, 2005	In 2 weeks of talks, delegates to the UN Climate Change Conference in Montreal concluded the decade-long round of negotiations that launched the Kyoto Protocol and opened a new round of talks to begin considering the future of the international climate effort.
COP-20 Lima, Peru December 1–12, 2014	All pledged to capitalize the GCF past an initial US$10 billion target. Multilateral assessment made developed countries come clean on emissions target. Agreed to put climate change into school curricula and climate awareness into national development plans.	COP-10 Buenos Aires, Argentina, December 6–17, 2004	Assessed progress and considered the challenges ahead, including efforts by developing countries to accelerate assistance for adaptation to climate change, the treatment of sinks in the Protocol's CDM, and the treatment of emissions from marine and aviation fuels.
COP-19 Warsaw, Poland, November 11–23, 2013	Developed countries rejected calls to set target of aid until 2020 when 100 billion US$/year will be available for developing countries. Agreed to a new "Warsaw International Mechanism" to provide expertise and aid to help developing nations cope with losses from extreme events related to climate change. Under the CDM, the conference agreed to help ailing mechanism, encouraging countries without legally binding emissions targets to use carbon credits called Certified Emission Reductions. The conference agreed on a multibillion-dollar framework to tackle deforestation through GCF.	COP-9 Milan, Italy, December 1–12, 2003	Adopted the IPCC Report on changes in carbon concentrations resulting from land-use changes and forestry. Extended the CDM by including modalities and scope for carbon-absorbing forest management projects. Also proposed to develop two funds: Special Climate Change Fund and Least Developed Countries Fund. Took 22 decisions, passed 1 resolution.

(*Continued*)

Table 3.1 Tabulated History of International Negotiations (*cont.*)

COP	Discussion and Outcome		Discussion and Outcome
COP-18 Doha, Qatar, November 26–December 7, 2012	Set out a timetable to adopt a universal climate agreement by 2015, to come into effect in 2020. Saw the launch of a Second Commitment Period under the Kyoto Protocol from January 1, 2013, to December 31, 2020. Adopted the Doha Amendment to the Kyoto Protocol. Took 26 decisions, passed 1 resolution.	COP-8 New Delhi, India, October 23–November 1, 2002	The Delhi Declaration reiterated the need to build on the outcomes of the World Summit, reaffirming that economic and social development and poverty eradication are the first and overriding priorities of developing country Parties to the Convention. Took 25 decisions, passed 1 resolution.
COP-17 Durban, South Africa, November 28–December 9, 2011	Adopted a universal climate agreement by 2015. Also agreed on a second commitment period of the Kyoto Protocol from January 1, 2013. A significantly advanced framework for the reporting of emission reductions for both developed and developing countries was also agreed. Took 19 decisions, passed 1 resolution.	COP-7 Marrakech, Morocco, October 29–November 9, 2001	Parties agreed on a package deal, to ensure compliance with commitments, consideration of LULUCF Principles in reporting of such data, and limited banking of units generated by sinks under the CDM. Took 39 decisions, passed 2 resolutions.
COP-16 Cancún, Mexico, November 29–December 10, 2010	The Cancún Agreement introduced mitigation pledges and operational elements such as a new GCF for developing countries and a system of "international consultations and analysis" to help verify countries' actions. Agreement hinged on finding a way to finesse the more difficult questions of if, when, and in what form countries will take binding commitments. The final outcome leaves all options on the table and sets no clear path toward a binding agreement.	COP-6 Bonn, Germany, July 16–27, 2000 The Hague, Netherlands, November 13–24, 2000	Consensus reached on Bonn Mandate. Work was also completed on a number of detailed decisions based on the Bonn Agreements, including capacity building for developing countries and countries with economies in transition. Decisions on several issues, notably the mechanisms LULUCF and compliance, remained outstanding. Took 6 decisions, passed 3 resolutions.
COP-15 Copenhagen, Denmark, December 7–18, 2009	A new political accord struck to provide explicit emission pledges by all the major economies—including, for the first time, China and other major developing countries—but charts no clear path toward a treaty with binding commitments.	COP-5 Bonn, Germany, October 25–November 5, 1999	A focus on the adoption of the guidelines for the preparation of national communications by Annex I countries, capacity building, transfer of technology and flexible mechanisms. Took 22 decisions.

COP-14 Poznan, Poland, December 1–12, 2008	Governments resolved to shift into "full negotiating mode" in hopes of delivering a comprehensive new climate change agreement in December 2009 in Copenhagen.	COP-4 Buenos Aires, Argentina, November 2–14, 1998	The Buenos Aires Plan of Action focusing on strengthening the financial mechanism, the development and transfer of technologies, and maintaining the momentum in relation to the Kyoto Protocol was adopted. Took 19 decisions, passed 2 resolutions.
COP-13 Bali, Indonesia, December 3–15, 2007	In intense and chaotic talks that ran a full day longer than planned, delegates remained far apart on fundamental issues but in the end agreed to launch a loosely framed negotiating process with the ambitious goal of achieving a new global climate agreement in 2009.	COP-3 Kyoto, Japan, December 1–11, 1997	The Kyoto Protocol was adopted by consensus. The Protocol includes legally binding emissions targets for developed country (Annex I) Parties for the six major GHGs, which are to be reached by the period 2008–12. Issues for future international consideration include developing rules for emissions trading, and methodological work in relation to forest sinks. Took 18 decisions, passed 1 resolution.
COP-12/MOP 2 Nairobi, Kenya, November 6–17, 2006	Continued discussion on two processes launched the preceding year in Montreal to consider next steps in the international climate effort, and agreed in the final hours to open another track to review the Kyoto Protocol.	COP-2 Geneva, Switzerland, July 8–19, 1996	A decision on guidelines for the national communications to be prepared by developing countries was adopted. Also discussed: QELROs for different parties and an acceleration of the Berlin Mandate. Took 17 decisions, passed 1 resolution.
Kyoto Protocol Kyoto, Japan, 1997–2012	This Protocol sets binding targets to reduce emissions 5.2% below 1990 levels by 2012. The Protocol made the emissions targets binding legal commitments for those industrialized countries that ratified it (the United States did not ratify it). The Kyoto Protocol brought out the terms CDM and carbon credit. Entered into force February 16, 2005.	COP-1 Berlin, Germany, March 28–April 7, 1995	Commitments were inadequate to meet the objective of UNFCCC. In a decision known as the Berlin Mandate the commitments of developed countries were sought. Took 21 decisions, passed 1 resolution.

(*Continued*)

Table 3.1 Tabulated History of International Negotiations (*cont.*)

COP	Discussion and Outcome		Discussion and Outcome
IPCC Established by UNEP and WMO in 1988	The leading international body for the global assessment of climate change. It does not conduct independent research; rather, it convenes climate experts from around the world every 5–7 years to synthesize the latest climate research findings in peer-reviewed and published scientific/technical literature. The IPCC issued comprehensive assessments in 1990, 1995, 2001, 2007, and 2014. IPCC reports are never policy prescriptive but the conclusions are relevant to nations, states, and businesses interested in enacting policies to limit future warming and reduce the costs of climate change.	UNFCCC 1992, Rio Earth Summit, Brazil	Aimed to stabilize GHG concentrations in the atmosphere "at a level that would prevent dangerous anthropogenic interference with the climate system." It also set a voluntary goal to reduce emissions from developed countries to 1990 levels by 2000—a goal that most countries did not meet. By 2015, 191 parties, including the US, have ratified the UNFCCC.

1 carbon credit = 1 tonne of = 1.1023 ton CO_2. This can be traded. Any nation may earn the right to emit 1 tonne of CO_2 if it funds projects aimed at reducing CO_2/GHG elsewhere by the same amount. Similarly, a nation will be entitled to receive funds, grants, or CDM facilities if it reduces its emissions by 1 tonne of CO_2.
CDM, *Clean Development Mechanism;* COP, *Conference of Parties;* GCF, *Green Climate Fund;* GHG, *Greenhouse Gas;* IPCC, *Intergovernmental Panel on Climate Change;* LULUCF, *Land-Use Change and Forestry;* QELROs, *Quantified Emissions Limitation and Reduction Objectives;* UNFCCC, *United Nations Framework Convention on Climate Change.*

Table 3.2 UNFCCC Negotiations: Who Stands Where Vis-à-Vis South Asia

Developed World	Developing World	South Asia
Carbon pricing		
A uniform global price is suggested to provide an incentive for clean technology.	Prices have to be rationally determined for each nation to avoid top-down approach. Rich and poor nations cannot have uniform pricing.	India has a domestic PAT system to price and regulate carbon emissions. However, price cannot be used as a tool to dampen country's economic growth.
Mitigation		
Energy economics must include emission reduction or capping to obtain best results.	Only rich nations are responsible for climate change. They must reduce emissions first. For others emissions control should be voluntary.	Nations must be given choice to commit for reduction of carbon emissions on a voluntary basis. However, all should ensure per capita emissions must not rise above global average.
Adaptation		
Quantification of loss and damage is very difficult.	Focus on mitigation and adaptation on a long-term basis.	Climate finance may be equally divided between mitigation and adaptation.
Technology transfer		
Joint research through collaboration is possible, but not at the cost of patent regime.	Technology must be transferred at low cost to poor and vulnerable nations.	Technology transfer is a must without disturbing patent regime, and cost should not get in the way.
Climate financing		
Committed to provide funds to poor nations from their government, privately, and through other development assistance funds.	The entire funding and finance have to come from the government exchequers of the rich nations. Money from assistance program must not be siphoned off.	Support money from rich nations must come from their government coffers, not from other assistance programs.
Transparency		
A universal review and reporting system for all nations must be worked out for annual auditing.	Wants more rigorous review mechanism for the developed world. Institute an open review system.	Review should be nationally defined and intrusive in nature. Questions can be asked only for projects funded by rich nations.

(Continued)

Table 3.2 UNFCCC Negotiations: Who Stands Where Vis-à-Vis South Asia (*cont.*)

Developed World	Developing World	South Asia
Differentiating responsibility		
Withdraw terms like industrialized nations and developing countries, Kyoto Protocol states. All nations should be brought under a universal agreement.	Seeks to continue with term "Kyoto Protocol states." Rich nations are to be mandated to reduce carbon emissions and other nations are to do so voluntarily.	Differentiation must continue based on equity and capability as per national circumstances.
Legal charter		
Rich nations, except USA and Canada, want COP-21 to be binding for all countries. USA wants commitment to reduce carbon emissions on a voluntary basis.	Differentiation must continue. Do not try to change the basic principle of UN climate convention.	While continue to support the stand of the developing world, India wants agreement to be equitable.
Intention		
All countries showed voluntary intention (INDC) to reduce GHG emissions by 25%–30% per unit of GDP of 2005 level by 2025–30 at the latest.		

PAT, *Performance achievement and trade.*
Source: Modified after UNFCCC (2017).

The insistence of first-world countries to remove an estimate of the amount of money necessary to be paid to second and third worlds for adaptation measures (US$100 billion a year) brought out the differences open. The first world, instead of "aid", insisted on cooperation between the countries to develop specific technology and cited example of India–USA joint cooperation on clean energy development. Scaling up of the Indo-US cooperation to South Asia–US cooperation may be worked out.

The lapse of the Kyoto Protocol in 2012 proved costly, as the climate scientist forecasted that a peak in emission of CO_2 and other GHGs by 2015 would cause more droughts, floods, mudslides, and rising seas. However in an alternative finding, the Smithsonian Tropical Research Institute (STRI) found that the world's forest would not perish but could adapt to and even flourish future climate change more innovatively. This development could probably make the follow-up negotiations difficult (Jaramillo et al., 2010). This study found that plants adapt more efficiently with their water use when it became scarcer. Rather than global warming, the predicament for tropical plants is deforestation. The fossil record shows that the plants can deal better with high temperatures and CO_2 in the absence of humans. Amidst conflict of interest, economy replaced democracy, as negotiations shifted from the UN General Assembly to the various economic groups, such as IMF, World Bank, G8, G15, and BRICS.

3.3 PARIS AGREEMENT

One of the major far-reaching monumental deliberations on climate change, after the Rio Conference and Kyoto Protocol, was the 21st Conference of the Parties (COP-21). It was held in Paris from November 30 to December 11, 2015. With more than 40,000 delegates from 195 countries in attendance, it was one of the most important meetings, graced by top leaders of almost all countries. This provided a clear political will to the conference to reach an agreement. Because the Kyoto Protocol will be elapsed in 2020, it was imperative that the attendees leave Paris with a new agreement to begin by that time and enable long-term changes beyond that.

In this background, the Paris Conference was called to—(a) frame a universal, legally binding agreement that will enable the world to combat climate change effectively and boost the transition toward resilient, low-carbon societies and economies, (b) reduce the GHG emissions to limit global warming to below 2°C—leaving 80% of fossil fuels unexplored, and societies' adaptation to existing climate changes, (c) mobilize US$100 billion per year by developed countries, from public and private sources, from 2020 to enable developing and low-developed countries to combat climate change whilst promoting fair and sustainable development (initial capital received US$10.2 billion, including almost US$1 billion from France), (d) guide economic and financial stakeholders toward redirecting their investments to launch the transition to low-carbon economies, and (e) suggest viable mechanism to maintain Carbon Budget, to reduce GHG by 40%–70% on 2010 levels by 2050, and to stop the atmosphere from reaching its maximum storage capacity of CO_2 by the 2030s (Miller, 2015).

The COP-21 marked a historic turning point in global action on climate change (Dagnet et al., 2016). On December 12, 2015, 195 countries adopted the Paris Agreement, a universal agreement that sets the world on a course to a zero-carbon, more resilient, and hopeful future. Building on the foundation of national climate plans, known as intended nationally determined contributions (INDCs) from 188 countries, the success of COP-21 and the Paris Agreement was a reflection of the desire of the nations and various civil society organizations, including faith-based communities, to save the blue planet. The Paris Agreement, which came into effect on November 4, 2016, received acceptance from almost all the countries in the world, as it was accommodating and had chosen a middle path among the warring three worlds. Unlike its predecessor, the Kyoto Protocol, which sets commitment targets that have legal force, the Paris Agreement, with its emphasis on consensus building, allows for voluntary and nationally determined targets. The targets are thus politically encouraged, rather than legally bound. Its action tool, the Sustainable Development Mechanism (SDM), will replace the Clean Development Mechanism (CDM) of the Kyoto Protocol to help the nations pursue emission reductions collaboratively.

The disagreements among the countries were largely solved while formulating the Paris Agreement. For example, keeping with the demands of small low-developed countries, COP-21 decided to reduce further the limit of temperature rise from 2°C to 1.5°C from preindustrial level. Similarly, keeping in mind the requirements of rapid developing nations (like India, Brazil, and South Africa), distribution of responsibilities to arrest climate change were made unambiguous. Although there was no specific mention (pointing finger) of historical responsibility (insisting on that would have vitiated the atmosphere), the developed countries have clearly agreed to accept the additional financial responsibility to stem the rot by agreeing to create a Green Fund of 100 billion US$ annually to support developing and low-developed countries to generate clean technology and mitigate climate change impact. Recognizing that arresting climate change is the responsibility of all nations in the world, COP-21 decided that starting from 2020, the UNFCCC will monitor and supervise the policy and implementation of all the countries in this regard. The countries should be reporting every 5 years to the UNFCCC, with the first evaluation starting in 2023.

The decision of the USA in 2017 to withdraw from the Paris Agreement was unfortunate. The Agreement is neither renegotiable, nor reversible. In contrast, France is planning to introduce a bill in its parliament to ban all new permits for fossil fuel exploration in France. Except USA, all other countries believe that the Paris Agreement is the way to bring about economic growth, create jobs, eradicate poverty, and enhance the quality of life of those who exist on this Earth, sustainably. Perhaps the biggest gain for mankind in COP-21 (2015) through COP-23 (2017) conferences is the manner in which the principle of "common but differentiated responsibilities" has been elucidated in the Paris Agreement. This was also a part of the 1992 UN Climate Convention mandate. At the end, it appears that none of the three worlds (developed, developing and least developed countries) have won in Paris. All have compromised their positions, and rightly did so. But if somebody has really won in Paris it is the "possibility" that this blue planet could still be saved if all the countries come together to see the reasons, voluntarily.

3.4 FUTURE STRATEGY AND SOUTH ASIA

It seems now that climate policies discussed so far have less to do with the climate and the environmental protection, and more to do with economic trading and commerce (Liberty Institute, 2002). The economic growth, till date, has gone hand in hand with the increase in GHG emissions from oil, gas, and coal. One percent growth means one percent more emissions and to be affluent has to burn coal, oil or gas.

For example, coal is still economical and is available in abundance in countries such as India and China. Naturally, these countries would not like to decouple development from using coal (and emission from coal). While the developed countries talk about environment, the developing countries are busy evolving. As an example for South Asia, the problem was how to get 1000 million poor people of this region in to the middle income class. Whether there was a coal power plant or whether the labor standards in the coal mines were low was second priority, as it was with the developed countries in the 19th Century. The developing countries believe that it is an old trick of the developed world to shift the goalpost of progress as soon as the developing countries reach the level of progress set by first world. People are realizing that causes of climate change lie in the north and the consequences in the south. One must not forget the historical responsibility of developed countries in combating the climate change.

In South Asia, India might face a huge influx of climate-refugees from Afghanistan, Pakistan, and Bangladesh, should high degrees of climate change occur. Drought and flooding linked to sea level rise would place the governments of these countries under severe stress and may lead to large-scale migration. In addition, this would combine with an internal population shifting from rural to urban areas, further increasing demographic pressure in cities. Moreover, long-term water stress in South Asia could become so severe that all previous agreements over water-resource sharing between India, Pakistan, China, and Bangladesh could break. The simulations also raise the possibility that, if current trends in emissions of SO_2 and black carbon continue, the drought frequency in South Asia may increase by double in the coming decades (Ramanathan et al., 2005).

The way global negotiations are progressing it appears that financial institutions would be interested to invest, trade, and develop various fiscal instruments, such as insurance, rate of interest, brokerage and so on, from which they could make profit through speculative positions on carbon prices. While environmental activists will continue to take ideological stand on GWCC favoring strict ethical and legal controls on emission, the hydrocarbon producing agencies may not support such strict regulations. The renewable energy producers would, however, welcome any strict directives against fossil fuel companies in the form of collecting taxes directly or through a carbon trading mechanism. Governments would also have the option to collect increased taxes from carbon emitting industries.

Although not squarely expected for the fast growing South Asian industries and urbanization effort to drastically reduce GHG discharge, this region must work innovatively to trim down such emissions. Being the regional leader, India has emerged as South Asia's principal spokesman on climate change. Hence, India must work in tandem cooperation to other South Asian nations to design a unified regional approach. India partnered with BRICS nations (Brazil, Russia, China, and South Africa) to represent the developing world in Copenhagen-COP15-2009, Cancun, Durban, Doha, Warsaw, and Lima-COP20-2014 summits. However, the modification in her approach to speak for the entire South Asia in Paris-COP21-2015, Marrakech, and Bonn-COP23-2017 was certainly appreciable.

As the climate change and sea level rise are going to impact South Asia most, SAARC must pursue a united face to the world to succeed at the global platform. South Asia must also stress on modifying lifestyles of its people to reduce emissions, and deploying renewable energy equipment to generate power. In addition, rational education and appropriate counselling of the people will be of great help to bring about a change in their mindset to live an ethics based responsible lifestyle.

CHAPTER 4

Carbon Policy of South Asia

Constituting close to one-fifth of the world population, South Asia is one of the most vulnerable regions in the world to the impacts of climate change. Earlier, Chapter 2 showed how the high accumulation of GHG in the atmosphere of South Asia and the World is impacting cyclone frequency, rainfall pattern, agricultural yield, health issues, food security, ecosystem, sea level fluctuations, and the change in coastal shoreline. The related carbon politics and what that mean to South Asia is discussed in chapter 3 hence, in this chapter, the existing climate policies of the eight South Asian nations are glossed over to understand where they stand as of today. While offering a brief account of their implementation, the shortcomings, if any, in the implementation of such policies were also explored in this chapter.

4.1 POLICY AND STRATEGY

The entire South Asia (also known as the Indian subcontinent) is a distinctly separate geological identity. It separated from the Gondwanaland by around 90 Ma and moved more than 3000 km north-northeast wards at variable speed (3–20 cm/year), till it crashed into the Eurasian Plate by around 51 Ma to form the mighty Himalayas (Mukhopadhyay et al., 1997, 2012). Currently, the Indian Plate is moving northeast at 5 cm/year, while the Eurasian Plate is moving north at only 2 cm/year. This is causing the Eurasian Plate to deform, and the Indian Plate to compress.

The overall assessment of climate change data suggests that the entire South Asia is climate vulnerable (IPCC, 2014). The coastal states of Bangladesh, India, Pakistan, and Sri Lanka are especially threatened by rising sea levels and flooding. The small but densely populated Maldives—the lowest-lying country in the world—faces the very real threat of inundation. Even the landlocked countries in South Asia like Afghanistan, Bhutan, and Nepal face threats from rising temperatures, drought, and glacial melt. The overall impacts of climate change on South Asia are summarized in Table 4.1 from discussion made in

Climate Change. http://dx.doi.org/10.1016/B978-0-12-812164-1.00004-9

Table 4.1 Summary of Impact of Climate Change in South Asia

Sl	Impact Summary
1	**Enhanced temperatures** More CO_2 in atmosphere means more growth of plants, but also more demand on water. Rain-fed wheat grown at 450 ppm CO_2 demonstrated yield increases with temperature increases of up to 0.8°C, but declines with temperature increases beyond 1.5°C (Xiao et al., 2005). Again, in Pakistan wheat yields are predicted to decline by 6%–9% in subhumid, semiarid, and arid areas with 1°C increase in temperature (Sultana and Ali, 2006), while even a 0.3°C decadal rise could have a severe impact on important cash crops like cotton, mangoes, and sugarcane. In Sri Lanka, half a degree temperature rise is predicted to reduce rice output by 6%, and increased dryness will adversely affect yields of key products like tea, rubber, and coconut (MENR Sri Lanka, 2000). In the hot climate of Pakistan, an increase of 2.5°C in temperature would translate into much higher ambient temperatures in the wheat planting and growing stages.
2	**Rainfall and water management** Water resources are inextricably linked with climate. Tendencies of increase in intense rainfall with the potential for heavy rainfall events spread over a few days are likely to impact water recharge rates and soil moisture conditions. Of 2,323 glacial lakes in Nepal, 20 have been found to be potentially dangerous with respect to GLOFs. The most significant such event occurred in 1985, when a glacial lake outburst caused a 10–15 m high surge of water and debris to flood down the Bhote Koshi and Dudh Koshi rivers for 90 km, destroying the Namche Small Hydro Project (Raut, 2006). Rapid depletion of water resource is already a cause for concern in many countries in South Asia. About 2.5 billion people will be affected with water stress and scarcity by the year 2050 in South Asia (HDR, 2006). With the islands of the Maldives being low-lying, the rise in sea levels is likely to force saltwater into the freshwater lens. The groundwater is recharged through rainfall. Although the amount of rainfall is predicted to increase under an enhanced climatic regime, the spatial and temporal distribution in rainfall pattern is not clear (Ministry of Environment and Construction Maldives, 2005).
3	**Extreme events** South Asia suffers an exceptionally high number of natural disasters. Between 1990 and 2008, more than 750 million people—50% of the region's population—were affected by a natural disaster, leaving almost 60,000 dead and resulting in about $45 billion in damages. Several studies showed that generally the frequency of occurrence of more intense rainfall events in many parts of South Asia has increased, causing severe floods, landslides, and debris and mud flows, while the number of rainy days and total annual amount of precipitation has decreased (Lal, 2003; Mirza, 2002). An increase in the frequency of droughts and extreme rainfall events could result in a decline in tea yield, which would be the greatest in regions below 600 m (Wijeratne, 1996). With the tea industry in Sri Lanka being a major source of foreign exchange and a significant source of income for laborers, the impacts are likely to be grave. On an average during the period 1962–88, Bangladesh lost about 0.5 million tonnes of rice annually as a result of floods, accounting for nearly 30% of the country's average annual food grain imports (Paul and Rashid, 1993). Again, excessive water withdrawals can exacerbate the impact of drought. Changes in the frequencies of extreme events will have an impact on land degradation processes such as floods and mass movements, soil erosion by both water and wind, and soil salinization.

Table 4.1 Summary of Impact of Climate Change in South Asia (*cont.*)

Sl	Impact Summary
4	**Crop yield, agricultural productivity** Agriculture is the mainstay of several economies in South Asia. It is also the largest source of employment. The sector continues to be the single largest contributor to the GDP in the region. As three-fifths of the cropped area is rain-fed, the economy of South Asia hinges critically on the annual success of the monsoons (Kelkar and Bhadwal, 2007). In South Asia, there could be a significant decrease in nonirrigated wheat and rice yields for a temperature increase of greater than 2.5°C, which could incur a loss in farm-level net revenue of 9%–25%. One study points out that in Bangladesh production of rice and wheat might drop by 8% and 32%, respectively, by the year 2050. Studies show that a 0.5°C rise in winter temperature could reduce wheat yield by 0.45 tons per hectare in India. Other studies suggest that 2%–5% decrease in Indian wheat and maize yield potentials for temperature increases of 0.5–1.5°C could occur. For countries in South Asia, the net cereal production is projected to decline at least between 4% and 10% by the end of this century under the most conservative climate change scenario. In Sri Lanka, Somaratne and Dhanapala (1996) estimate a decrease in tropical rain forest of 2%–11% and an increase in tropical dry forest of 7%–8%. This study also indicates that increased temperature and rainfall would result in a northward shift of tropical wet forest into areas currently occupied by tropical dry forest. Droughts combined with deforestation increase fire danger (Laurance and Williamson, 2001). Global warming and climate change (GWCC) will also impact crop pests and diseases.
5	**Fisheries** A strong ENSO will decline in fish larvae abundance in the coastal waters of South Asia. There is a potential to substantially alter fish breeding habitats and fish food supply and therefore the abundance of fish populations in Asian waters due to the response to future climate change to the following factors: ocean currents, sea level, seawater temperature, salinity, wind speed and direction, strength of upwelling, the mixing layer thickness, and predator response.
6	**Sea level rise** Low-lying mega coastal cities will be at the forefront of impacts (Karachi, Mumbai, Chennai, Male, and Dhaka) and vulnerable to the risks of sea level rise and storms. A rise in sea level would raise the water table, further reducing drainage in coastal areas. All these effects could have possibly devastating socioeconomic implications, particularly for infrastructure in low-lying deltaic areas. The population of the Maldives mainly depends on groundwater and rainwater as a source of freshwater. Both of these sources of water are vulnerable to changes in the climate and sea level rise. With the islands of the Maldives being low-lying, the rise in sea levels is likely to force saltwater into the freshwater lens (Ministry of Environment and Construction, 2005).

Source: *Modified from IPCC, 2014. Climate Change 2014: Impacts, Adaptation, and Vulnerability. Part A: Global and Sectoral Aspects. In: Field, C.B., Barros, V.R., Dokken, D.J., Mach, K.J., Mastrandrea, M.D., Bilir, T.E., Chatterjee, M., Ebi, K.L., Estrada, Y.O., Genova, R.C., Girma, B., Kissel, E.S., Levy, A.N., MacCracken, S., Mastrandrea, P.R., White, L.L. (Eds.), Contribution of Working Group II to the Fifth Assessment Report of the Intergovernmental Panel on Climate Change. Cambridge University Press, Cambridge, UK, New York, NY, USA, p. 1132.; Sivakumar, M.V.K., Stefanski, R., 2011. Climate change and food security in South Asia. In Lal, R., Sivakumar, M.V.K., Faiz, M.A., Mustafizur Rehman, A.H.F., Islam, K.R. (Eds.), Climate Change and Food security in South Asia, Springer Science, p. 600; UNFCCC (2015).*

earlier chapters. These remained the base for the member countries to design their respective public policies on climate mitigation and adaptations.

However, the major problem faced by most of these countries is that in many situations the rational science-based doctrines have been compromised owing to political and social considerations.

4.1.1 Afghanistan

Afghanistan is a landlocked country encompassing an area of about 652,000 km^2 characterized by high mountains and deep valleys (Fig. 1.4). Around 63% of the country is mountainous, while the large flat terrain in the south-western part of the country is formed by the Helmand River's drainage basin. It has an estimated population of 28.6 million (2015) and a per capita GDP of US$660 (IMF, 2015). Despite suffering decades of political instability and war, Afghanistan has made considerable progress in recent years in terms of economy and infrastructure. Economically, about 85% of the Afghan population relies directly or indirectly on agriculture as their main livelihood, which constitutes about 28% of its GDP. The service sector contributes another 28%, while industry contributes around 20%. However, since the majority of agriculture in Afghanistan is rain-fed, change in rainfall pattern owing to transmuting climate, the majority of Afghans remain vulnerable.

Afghanistan ranks 169th out of 187 countries in the United Nations Development Program's Human Development Report 2014 and is listed under the lowest human development category. More than half of Afghans live below the poverty line, and the country has one of the lowest life expectancies at birth (60 years) in Asia. Moreover, only 46% of the population have access to safe drinking water, and the percentage of the population with electricity is around 30%, which is among the lowest in the world. However, there has been progress in human development indicators since 2002, with school enrolment increasing from 1 million to 8.6 million. Girls' school enrolment has also increased substantially to more than 3.6 million).

The country has a semiarid to arid climate, and influenced mostly by the monsoon and the cold northerly wind. Climatically, the mean annual temperature in Afghanistan has increased by 0.6°C since 1960, while the average annual precipitation has decreased by 0.5 mm per decade during the same period. Actually, the frequency of warm days and warm nights per year has increased by 25 and 26 days since 1960, while the frequency of cold days and cold nights per year has decreased by 12 days and 12 nights during the same period. Although the monsoon brings rainfall to the eastern provinces in the summer months, many areas in the western and southern Afghanistan are frequently associated with intense heat, drought, and sand storms.

The GHG emissions by Afghanistan are extremely low, 0.3 tons per capita, making Afghanistan one of the lowest emitters globally. The per capita energy consumption is also low at an average of 120kW/year. This situation is however likely to change with infrastructure development and rapid urbanization in the years to come. The most significant sources of GHG are the agriculture sector (53%), followed by land use change, deforestation (33%), and the energy sector (13%, Afghanistan Government Portal, 2015, Table 4.2).

In this background, efficient land management, designing a disaster management strategy, initiating research into climate changes and early warning systems, improving food security, and diversification of livelihoods are the prime areas where Afghanistan needs to progress (SAARC, 2014). In addition, lack of expertise, lack of reliable historical climate data, weak public awareness about environmental issues, and lastly unstable national security have remained areas of concern. Amid these shortcomings, the country has identified 10 key actions as part of its futuristic national adaptation plan (NAP) and estimated a cost of US$10.785 billion over the next 10 years (Afghanistan Government Portal, 2015). Being a member of the UNFCCC, Afghanistan has drafted in 2013 its initial national communication and presented its potential action plan toward mitigation and adaptation to climate change (Gorin and Clot, 2016). In fact, the country has a need to promote and strengthen adjustment strategies that aim at improving water management for agricultural practices and research.

In reality, the National Environmental Protection Agency (NEPA) of the Government of Afghanistan, in collaboration with United Nations Environment Program (UNEP), had launched the US$6 million initiative to protect war-torn communities vulnerable to the effects of vagaries of climate (NEPA 2013). This Global Environment Facility (GEF) funded initiative includes plantation of trees on terrace slopes in the Central Highlands, wherefrom most of the rivers originate, and support agricultural activities in the country (Fig. 1.4). This aforestation not only will mitigate drought but also could become a tourist destination.

The NEPA scheme is expected to be implemented in four locations: Badakhshan in the northeast, Balkh in the north, the area between Koh-e Baba and Bamyan, and Daikundi in the Central Highlands. The NEPA is also expected to help Afghan institutions to better deal with earthquake shocks and climatic hazards, and increase resilience at a decentralized level. NEPA's effort is further strengthened by Food & Agriculture Organization (FAO) and the World Food Program (WFP) supported by the United Nations Assistance Mission in Afghanistan (UNAMA). Such initiatives are expected to be accomplished with assistance from the USAID-funded Biodiversity Program of the Wildlife Conservation Society (WCS) and the local Afghan organizations and communities.

Table 4.2 Sector-Wise GHG Emissions in Afghanistan: Present and Future

Greenhouse Gas Emissions of CO_2, CH_4, and N_2O in Afghanistan in 2005–2030[a]							
	CO_2 Equivalent Gg				2020	2025	2030
GHG Emission Sector	CO_2	CH_4	N_2O	Aggregated	CO_2 eq, Gg	CO_2 eq, Gg	CO_2 eq, Gg
Energy	2,910.04	736.00	129.83	3,775.87	9,745.46	10,894.02	12,087.00
Industry	312.15	—	—	312.15	791.57	878.25	974.42
Agriculture	—	9,296.49	5,812.50	15,108.99	24,665.30[b]	29,578.77[b]	35,471.04[b]
Land use change and forestry	9,341.13	80.64	9.30	9,431.07	10,949.18	11,507.70	12,094.71
Waste	—	130.41	—	130.41	330.70[b]	366.91[b]	407.09[b]
Total GHG emissions, including LULUF	12,563.32	10,243.54	5,951.63	28,758.48	46,482.0	53,180.64	61,034.25
Total GHG emissions, excluding LULUF	3,222.19	10,162.90	5,942.33	19,327.42	35,533.02	41,672.95	48,939.54

[a]*Information used from ADB (2016), Afghanistan Greenhouse Gas Inventory Report and projection for 2020–2030 using GACMO model.*

[b]*CH_4 (CH_4 emission × 21) and N_2O (N_2O emission × 310) counted as CO_2 eq.*

Source: *Modified from Afghanistan Government Portal (2015); WRI (2015); FAO (2015).*

4.1.2 Bangladesh

Bangladesh is located in a low-lying delta, formed by the dense network of the distributaries of the mighty rivers: the Ganges, the Brahmaputra, and the Meghna (Fig. 1.4). The total land area is 147,570 km^2, with some hilly areas located in the northeast and southeast. A network of more than 230 major rivers and their tributaries crisscross the country. Bangladesh has a population of about 169 million and GDP of US$66.8 billion per annum (per capita US$3890). With over 1000 people per sq. km, the country has one of the highest population densities in the world. Although poverty has declined from 57% in 1990 to 40% in 2005, nearly 56 million people are still living below the poverty line.

The major climatic processes that influence Bangladesh have been easterly trade winds causing warm drier circulation, and the southwest monsoon bringing maximum rainfall between June and September. With an average elevation of only 4–5 m above the mean sea level, nearly a third of the country is susceptible to tidal inundation and nearly 70% gets flooded during heavy monsoons. About 10% of the country is only 1 m above the mean sea level, and one-third is under tidal excursions. In fact, about 60% of the worldwide deaths caused by cyclones during the past 20 years occurred in Bangladesh. The Bangladeshi economy is based predominantly on agriculture, forestry, and fishing, although recently it has diversified to include manufacturing. Agriculture, however, is still crucial as it supports a large number of people and most other sectors (like energy). Overall, the real threat comes from climate change which is expected to decrease agricultural GDP by 3.1% each year—a cumulative $36 billion in lost value-added for the period 2005–2050 (SAARC, 2014).

Bangladesh is the poster child for the potential impact of climate change. It is recognized by the UNFCCC as one of the most vulnerable countries to climate change impacts. The fact that in Bangladesh 1015 people reside in every km^2 (one of the most densely populated countries in the world), the climatic impact will be devastating. The increased intensity and frequency of flash floods under future climate change scenario is a major challenge for development, and a significant barrier to its vision of eliminating poverty and becoming a middle income country by 2021 (Nachmany et al., 2015).

Bangladesh, despite contributing only one-fifth of 1% of global CO_2, is going to be hit hard due to global warming (Table 4.3). In view of this Bangladesh is aiming to pursue a four-pronged approach by cooperating with other South Asian countries to ensure allocation of sufficient funds from the developed countries to (1) cover the cost for effective mitigation, (2) allocate funds to achieve poverty alleviation goals by innovative adaptation, (3) oversee technology transfer like Clean Development Mechanism, and (4) build expertise.

Table 4.3 Sector-Wise GHG Emissions in Bangladesh: Present

	Emissions by Sector $MtCO_2$ eq (2012)	Percent of Total Emissions (%)
Agriculture	74.75	39
Energy	62.37	33
Land use change and forestry	31.35	17
Waste	18.28	10
Industrial processes	3.12	2

Source: *Modified from USAID (2015); WRI (2015); FAO STAT (2015).*

Bangladesh's journey to the middle-income category, will require increasing GDP growth to 7.5%–8% per year based on the accelerated export and remittance. Growth will also need to be more inclusive through productive employment opportunities. To sustain accelerated and inclusive growth, Bangladesh needs to manage urbanization more effectively, which may increase the demand for energy (Khan and Shahjahan, 2014).

Despite being a small country, Bangladesh represents a large variation in ecological characteristics. In a bid to develop a low carbon society by 2025, the Bangladesh Climate Change Strategy and Action Plan (BCCSAP) is formulating a carbon policy flexible enough to stretch over time and space. It is proposed to measure carbon emission by using ExSS (Extended Snapshot) tool in the energy sector and by using the AFOLUB (Agriculture, Forestry and Other Land-use Bottom-up) mitigating model in nonenergy sector (Fig. 1.4; Jilani et al., 2012).

Adopting participatory mechanisms such as information dissemination, fairs, media, community congregation, mobile network, among others could be effective. Skill development and capacity building are also being made mandatory. Effort is made to integrate adaptation into planning process. Because of many social drawbacks, there can, however, be different adaptive measures or practices in different parts of Bangladesh. Many of such measures could be replicated elsewhere within the country and in the entire South Asia.

4.1.3 Bhutan

Bhutan with an area of 38,394 km^2 is a landlocked country located north of India and Bangladesh, and east of Nepal (Fig. 1.4). Bhutan is often visited by violent storms from the Himalayas, frequent landslides during the rainy season, and soil erosion. Limited access to potable water is also a major constraint of this small nation. Bhutan is home to at least 59 natural mountain lakes as well as some 2674 glacial lakes. Snowfalls are common above 3000 m height. However due to rising temperatures, glaciers in Bhutan are retreating at a rate of 30–40 m per year (IPCC, 2014, Table 2.4).

The climate is humid and subtropical in the southern plains and foothills, temperate in the inner Himalayan valleys of the southern and central regions, and cold in the north, with year-round snow on the main Himalayan summits. Temperatures vary according to elevation. Temperatures in the capital Thimphu, located at 2200 m above sea level in west-central Bhutan, range from approximately 15–26 °C during the monsoon season of June through September but drop to between about −4 and 16 °C in January. Most of the central portion of the country experiences a cool, temperate climate year-round. In the south, a hot, humid climate between 15 and 30 °C year-round, although temperatures sometimes reach 40 °C in the valleys during the summer. Annual precipitation ranges widely in various parts of the country: 40 mm in the north, 1000 mm in the central region, and about 7800 mm at some locations in the humid, subtropical south, ensuring the thick tropical forest, or savanna (SAARC, 2014).

Melting of glacier to form lakes and add water to rivers is a common feature in Bhutan. For example, Lake Imja Tsho while was virtually nonexistent in 1960, now covers nearly 1 km^2, as the Imja glacier has retreated at an unprecedented rate of 74 m per year between 2001 and 2006. Some other glacial lakes are increasing their size, and few by 800% over the past 40 years. The hydrological modeling of glacial lakes, terrain classification, and vulnerability assessment are important scientific means to understand GLOF (Glacial Lake Outburst Floods) impacts. The GLOF model helps in devising mitigation measures and early warning systems (Shrestha et al., 1999, 2012).

However, GLOF mitigation measures and early warning systems applied in the Nepal and Bhutan Himalayas are quite expensive and require much detailed field-work and maintenance. As an alternative, regular temporal monitoring of glacial lakes by RADAR satellite-based techniques to detect any changes and provide an early warning is being used currently (SAARC, 2014; Shrestha et al., 1999).

Lying within the snow-capped mountains, the small isolated kingdom of Bhutan showcases possibly a carbon negative economy (Munawar, 2016). Actually, against the emerging threats of climate change, Bhutan as early as in 2006 brought out the National Adaptation Program of Action (NAPA) to work out the appropriate mitigation and adaptation strategies. In the background of sector-wise GHG emission, NAPA identified eight priority projects, a brief introduction on that will be timely. The projects are:

EFRC-Capacity Building: Climate Resilient and Environment Friendly Road Construction was suggested to be taken up as a course curriculum in two national technical institutes of the country (College of Science and Technology and Jigme Namgyal Polytechnic) to develop home-grown engineering capacity in robust road making, control erosion, and increased safety for communication and transport facilities. This visionary project is going on smoothly.

Community-based Food Security: This project is to support the vulnerable communities in building their resilience to climate change through various adaptation measures, by (1) community mobilization, advocacy, and capacity building, (2) scaling up resilient food production and postharvest processing technologies at the household or community level, (3) increased food production, and (4) drudgery reduced making time available for other gainful employment. Implementation has been satisfactory.

Community-based Forest Fire Management and Prevention: This project is to enhance the capacity of rural communities on the management of forest fires through proper planning and using appropriate tools and equipments by (1) institutionalizing community level FFMP, (2) reducing the incidence of forest fire, (3) building capacity to develop and maintain advanced fire fighting equipments, and (4) contributing to the country's national goals and international commitments on GHG emission, apart from the reduction in loss of lives and properties.

Disaster Risk Reduction and Management: This project is set to (1) provide emergency and rapid high class medical services to vulnerable communities, (2) enable communities to manage adverse impacts of natural disasters, (3) enhance the capacity of health services, (4) form District Disaster Management Committee, and (5) conduct human resource training, orientation and sensitization in managing disasters.

Enhancing National Capacity for Weather and Flood forecasting: The project intends to enhance quality time series climate data, improve internet connection in remote villages, and strengthen technical and professional capacity building to acquire, process, and disseminate climate data in-house. However, land acquisition for installing Automatic Weather Station (AWS) has been a problem, and limited the availability of GSM (global system for mobile communication)/GPRS (general packet radio service)/Internet.

Flood Protection of Downstream Industrial area: This project is to predict, effectively intervene and to teach affected people adaptive measures during major landslide and flood. Innovative interventions to reduce the intensity of natural calamities are most sought after.

Managing Landslides and Preventing Flood: This project aims to (1) reduce the risk of landslides through various slope stabilization techniques, (2) prepare landslide hazard map, and (3) adopt appropriate methodology and train human resource to take remedial measures for major landslide in the area with respect to slope stability. However, the main constraints have been the lack of technical expertise on the mitigation techniques and the possibility of poor coordination among key stakeholders.

Rainwater Harvesting and Drought Adaptation: This project aims to (1) develop a water resource inventory through GIS (Geographic Information System) mapping of all water resources (entire river basins, their tributaries and the lakes, wetlands, streams), (2) adopt integrated water resources management system to address the holistic water resources management across the country, (3) examine site(s) for new reservoir construction, and (4) provide safe and adequate drinking water through rainwater harvesting. The project further aims to hold community awareness through celebration of World Environment Day, World Water Day, workshops, seminars, and community mobilization.

The above projects have started showing results. The energy and climate intelligence unit's carbon comparator tool has declared Bhutan to be an "unparalleled carbon sink, absorbing three times the amount of CO_2 than its population emits." Bhutan maintains carbon net negativity due to (1) low levels of industrial activity, (2) 100% electricity generation through hydropower, and (3) 70% of land covered by forests (Munawar, 2016). Bhutan, thus, showcases to the world how a less developed country can improve its economy without increasing GHG emissions.

4.1.4 India

With a land area of 3,287,264 km^2 (90% land), India has a coastline of 7516.6 km. India can be divided into five physiographic regions (Fig. 1.4): Northern Mountains, Indo-Gangetic Plains, Peninsular Plateaus, Islands, and Coastal Plains. Climatically, India ranges from arid desert in the west, alpine-tundra and glaciers in the north, humid tropical regions supporting rainforests in the southwest and the island territories, and equatorial in far south. The country displays four seasons: winter (December–February), summer (March–May), a rainy monsoon season (June–September), and a postmonsoon period (October–November). Summer lasts between March and June in most parts of India, where temperatures can exceed 45°C during the day. The highest temperature recorded in India was 50.6°C in Alwar (Rajasthan) during 1956, and the lowest was −45°C in Drass (Jammu & Kashmir). The coastal regions exceed 30°C coupled with high levels of humidity. The temperatures in the Thar Desert area in Rajasthan normally record the maximum. The Himalayas act as a barrier to the frigid wind flowing down from Central Asia. Thus, north India is kept warm or only mildly cooled during the winter; in summer, the same phenomenon makes India relatively hot.

The rain-bearing monsoon clouds are attracted to the low-pressure system created by the Thar Desert. The southwest monsoon splits into two arms the Bay of Bengal in the east and the Arabian Sea in the west, and is responsible for agroeconomic activities of the entire South Asia. Winters in peninsular India see mild to warm days and cool nights. Further north, the temperature is cooler.

Temperature in some parts of the Indian plains sometimes freezes during the winter. Most parts of the northern India remain covered by fog during this season.

The recent trend shows that the mean annual temperature in India has increased by 0.56°C between 1901 and 2007, while seasonal mean rainfall has decreased. The sea level rose by 0.21 m as of 2009 since 1901. India's climate is influenced by the presence of the Himalayas in the north and the Thar Desert in the west. The northern part of the country is characterized by a continental climate with hot summers and cold winters. The coastal regions of the country, however, experience warmer temperatures with little variation throughout the year and frequent rainfall. The two most influencing climatic factors have been monsoon accounting for 80% of annual precipitation in India. The ENSO (El Nino Southern Oscillations) years are generally associated with less Indian summer monsoon rainfall, while La Niña years are associated with higher monsoon rainfall (SAARC, 2014).

The GHG emission in various activity-sectors in India is shown in Table 4.4. Within the climate change mitigation policy framework, India has taken several proactive steps (both domestically and internationally) to move beyond the GDP approach to incorporate Low Carbon Inclusive Growth (LCIG) pathways. Given the growth and consequent energy trajectory of India (from 549 Mtoe

Table 4.4 Sector-Wise GHG Emissions in India: Present and Future

Sector-Wise Emissions in Indian Cities	(%)
Residential	28
Transport	24
Industrial	33
Commercial	10
Others	4
Waste	1

Emission Inventory Projections

Gases (MT)	1995	2005	2015	2025	2035	CAGR[a]
Carbon	212	333	492	646	738	3.1
Methane	17.6	19.5	21.5	23.2	25.7	0.9
N_2O	0.2	0.3	0.5	0.7	0.8	3.5
CO_2 eq	1219	1726	2413	3075	3504	2.7
SO_2	4.8	5.6	7.4	8.4	7.4	1.1
NO_x	4.1	5.6	6.9	8.2	8.7	1.9
Particulate	3.1	4.3	4.3	4	3	−0.1
CO	37.1	40.8	41.5	43.4	43.5	0.40

[a]*Compounded annual growth rate over 1995–2035 (%).*

Source: *Sector-wise emissions in Indian cities (from Sridhar, 2010) and Emission inventory projections (from Shukla et al., 2001).*

in 2011 to 1460 Mtoe in 2031, *million tons oil equivalent*), consumption of all types' of energy sources will continue to increase. And in the next few decades, there is no respite from using fossil fuels, with more than 70% of primary energy coming from fossil-based fuels mainly coal. Also the percentage share of imports of fossil fuels would also go up to 91% for oil, 66% for coal, and 61% for natural gas (TERI, 2015).

The first National Action Plan on Climate Change (NAPCC) of India was rolled out on June 30, 2008 (NAPCC 2008). The Plan, while focusing on eight national missions to be accomplished by 2017, outlines existing and future policies and programs dealing with mitigation and adaptation. The eight missions have been identified in the backdrop of overriding priority of maintaining high economic growth rates to push up the living standard of its poor people. The NAPCC, however, pledged that at no point GHG emission of India will exceed that of the developed nations, even when it pursues the development objectives. The eight climate mitigation missions of India are:

National Solar Mission aims to (1) promote development and use of solar energy for power generation in urban areas, industry and commercial establishment to ultimately make it competitive with fossil-based energy options, (2) increase production and deployment of photovoltaic to generate 1000 MW/year, (3) strengthen domestic manufacturing capacity, and (4) establish Solar Research Centre and collaborate with other countries in technology development.

National Mission for Enhanced Energy Efficiency will ensure (1) that big companies reduce energy consumption through trading energy-saving certificates, (2) energy-audit, (3) tax rebate/tax benefit to energy-efficient appliances, and (4) public–private partnership.

National Mission for Sustainable Habitat plans to (1) promote energy efficiency as a core component of urban planning, (2) ensure recycling of urban waste including generating power from it, (3) enforce fuel and exhaust economy standards strictly and offer tax rebate on purchase of efficient vehicles, and (4) offer an incentive to use public transportation.

National Water Mission sets a target of 20% improvement in water use efficiency through pricing and other measures.

National Mission for Sustaining the Himalayan Ecosystem will work to conserve biodiversity, forest cover, and other ecological values in the Himalayan region to hold back the retreat and melting of glaciers, the major water source of Indian rivers.

National Mission for a Green India commits aforestation of 6 million hectares of degraded forest land to expand forest cover by 10% as quickly as possible (from 23% to 33% of India's land area).

National Mission for Sustainable Agriculture aims to develop climate-resilient crops, expand weather insurance mechanism, and modify agricultural practices.

National Mission on Strategic Knowledge for Climate Change will (1) establish Climate Science Research Fund and collaborate with international agencies to support research programs that aims to improve climate modeling, (2) encourage private sector to develop adaptation and mitigation technologies through venture capital, and (3) develop supercritical technologies to generate power from renewable energy.

A mid-term review to this ambitious NAPCC mission was done after 4 years in 2012 (Byravan and Rajan, 2012). The review noted that although not much headway had been made work-wise, effort could still be made to sharpen the policies further. For example, National Solar Mission could take a leaf from the 1990s successful mission of the Ministry of Non-conventional and Renewable Energy in Maharashtra and Andhra Pradesh. Similarly, the National Mission for Enhanced Energy Efficiency may focus also on smaller manufacturing units, which may be polluting more than larger units. Likewise, the National Mission for Sustainable Habitat must focus on poor and vulnerable while planning for urban housing and transportation. The prescribed policies of National Water Mission could be made more integrated to climate change, while that under National Mission for Sustaining the Himalayan Ecosystem should focus on encouraging community to lead campaign on impact of forest degradation and dam spillage. The National Mission for Sustainable Agriculture could as well focus on the multidimensional impact of climate change on society, ethics, morality, and customs, while the pledge under the National Mission for a Green India appears to be too ambitious.

According to a UNEP-GRID report (2012), India emitted 1727.71 $MtCO_{2eq}$ in 2007, contributed by energy sector (58%), industry (22%), agriculture (17%), and waste (3%). This amounted to 1.5 tons of per capita emission (Table 4.4). The LULUCF (Land use, Land use change and Forestry) sector was a net sink, sequestering 177.03 $MtCO_{2eq}$. In this regard, it would not be out of place to mention the UNDP sponsored BERI Project (2010) from Karnataka, which aimed at developing and implementing a bioenergy technology package to reduce GHG emissions to promote a sustainable and participatory approach in meeting rural energy needs. The main components of the project were: (1) implement biomass gassifier for electricity generation, (2) establish community biogas systems for cooking and meeting domestic loads, and (3) integrate efficient community irrigation processes. A total of more than 2900 ha of forest and farm land have been cultivated since the inception of the BERI project.

Ramachandra and Shwetmala (2012), working on the decentralized carbon foot print analysis for opting climate change mitigation strategies in India, reviewed sector-wise carbon emission and sequestration inventories for all the states of India. They further stated that CO_2, CO, and CH_4 emissions from India are 876.49 Mt, 20.41 Mt, and 15.33 Mt per year respectively. Their analyses show that Maharashtra emits higher CO_2, followed by Andhra Pradesh, Uttar Pradesh, Gujarat, Tamil Nadu, and West Bengal. The carbon status, which is the ratio of annual carbon storage against carbon emission, for each state is computed. This shows that small states and union territories, where carbon sequestration is higher due to good vegetation cover, have carbon status >1. Annually 7.35% of the total carbon emissions get stored either in forest biomass or soil, out of which 34% in Arunachal Pradesh, Madhya Pradesh, Chattisgarh, and Orissa.

As per the UNFCCC-COP requirement, India in 2015 came out with its Intended Nationally Determined Contribution (INDC) keeping in view of its development agenda (COMEST 2010). This agenda includes the eradication of poverty, coupled with the commitment to follow the low carbon path using clean technologies. Under the INDC, India has voluntarily agreed to reduce the emissions intensity of its GDP by 20%–25% from 2005 levels by 2020 (INDC-India, 2015). The other commitments under INDC are to (1) propagate a healthy and sustainable way of living based on traditions and values of conservation and moderation, (2) reduce the emission intensity of its GDP to 35% by 2030 from the 2005 level, (3) use nonpolluting technology through Green Climate Fund (GCF) to generate about 40% of installed electric capacity from fuel based energy resources by 2030, (4) create an additional carbon sink of 2.5–3.0 billion tons of CO_2 equivalent through additional forest and tree cover by 2030, (5) enhance investments in sectors vulnerable to climate change, for example, agriculture, water, coasts, health, and disasters, and (6) build capacities; create domestic framework and international architecture for effective diffusion of cutting edge technology in India and South Asia.

It is further suggested that India is likely to meet or even exceed the INDC pledge based on its existing national policy missions and macroeconomic trends (Pahuja et al., 2014). Because India embraces one-sixth of world population and to meet the basic human and social developmental needs, an increase in GHG emissions, at least for few years may be required (TERI, 2015). Hence, India's response to climate mitigation will also be a response to its developmental changes.

India's threat from climate change will be acute as its nearly 40 million people will be vulnerable from rising sea levels. It also apprehends that higher temperatures could bring down agricultural productivity by 40% in the next 60 years in the country. India stands 10th in the list of highest per-capita emission countries in the world. Although this is significantly below the global average, the scenario might change in the decades to come as India would intend to double its power generation from her vast coal reserves by

2035. The subsurface burning of coal in many coal mines in the country emits GHG. India's coal production is however inhibited by a variety of factors—a large element of inefficiency, poor infrastructure, inadequate mining and beneficiation technology, and corruption in the coal mine block allotment and pilferage in transportation. Further, opaque procedures, archaic regulations, and policy uncertainties rate the ease of doing business low.

4.1.5 Maldives

Situated in the Indian Ocean, south of India, Maldives consists of approximately 1190 tiny coral islands grouped in a double chain of 26 atolls (Fig. 1.4). Most atolls consist of a large, ring-shaped coral reef supporting numerous small islands. Islands average only 1–2 km^2 in area, and lie between 1 and 1.5 m above the mean sea level. Although some of the larger atolls are approximately 50 km long from north to south, and 30 km wide from east to west, no individual island is longer than 8 km. On an average, each atoll has approximately 5– 10 inhabited islands.

The islands are spread over roughly 90,000 km^2, making this one of the most disparate countries in the world. Composed of live coral reefs and sand bars, the atolls are situated atop a submarine ridge (Chagos-Laccadive aseismic volcanic ridge created by eruptions from Reunion hotspot volcano during Paleocene-Eocene) 960 km long that rises abruptly from the depths of the Indian Ocean and runs from north to south. Only near the southern end of this natural coral barricade, two passages permit safe cruising of ship from one side of the Indian Ocean to the other through the territorial waters of Maldives. The largest island of Maldives is Gan, which belongs to Laamu Atoll. The presence of large South Asian subcontinental landmass in the north causes differential heating of land and water. In Maldives, the southwest monsoon lasts from the end of April to the end of October and brings the worst weather with strong winds, storms, and flooding damaging properties. The climate of Maldives is greatly influenced by its tropical monsoon weather, which lasts between May and October. The annual mean temperature is around 28°C (range 24–33°C) with little interseasonal variability, and annual rainfall averages 2540 mm in the north and 3810 mm in the south.

The southwest monsoon produces strong westerly winds in the country, the strongest winds being observed in the northernmost parts. The NE monsoon brings drier than average conditions particularly in the north and central regions, and produces strongest northeast winds in the same regions. Northeast winds are relatively weak or absent in the southern regions. The recent trend shows that the mean annual temperature has been increasing by 0.17°C and 0.07°C per decade since 1960 (SAARC, 2014).

Several autonomous bodies such as the Environment Research Centre, Maldives Energy Authority, and Maldives Water and Sanitation Authority under the Ministry of Environment, Energy and Water (MEEW) have drawn up the plan of mitigation and adaptation policy in the face of climate change and are spearheading the specific actions under well defined strategies and objectives.

Under worst-case scenario of climate change the entire Maldives would be submerged in water (IPCC, 2014). This fact led the then President of the country in 2009 to examine moving the entire population to a new sovereign territory, to be built by using tourist dollars. The tourism industry actually accounts for approximately 70% of GDP of Maldives. A "green tax" on tourists was first raised in 2009 at a rate of US$3 per day, which later revised in 2014 to US$6 per day per guest (Nachmany et al., 2015).

In the country's 2001 national report to the UNFCCC, Maldives volunteered to participate in international advocacy, incorporate climate change concerns into regulatory policies (*Strategic National Action Plan for Disaster Risk reduction*), create financial mechanisms to implement climate change programs, build capacity to adapt climate change, add adaptive measures into national planning, and develop strategies to mitigate GHG emissions (*Adaptation program 2010–2020;* Nachmany et al., 2015).

While Maldives will still generate about 3.3 million tons CO_2 equivalent by year 2030, she however intends to take effective actions to reduce unconditionally 10% of its GHG emissions by the same year. In fact, the country showed inclination to further reduce GHG emission even by 24% if international funds are made available (Marcu et al., 2015; MEEG-Maldives, 2015).

4.1.6 Nepal

Nepal is located north of India and extends along the Himalayan axis for a length of 800 km, with the widths varying from 150 to 250 km. Its 147,181 km^2 area is landlocked and show tremendous geographic diversity. It rises from as low as 59 m elevation in the tropical Terai to over 7000 m. The Earth's highest Mount Everest or Sagarmatha (8848 m) is located in Nepal and have about 90 smaller peaks (Table 4.5). In addition to the continuum from tropical warmth to cold comparable to Polar Regions, average annual precipitation varies from as little as 160 mm in the rain shadow north of the Himalaya to as much as 5500 mm on windward slopes in the Terai region (south of the Himalayas). Topographically, Nepal can be divided into four belts along a south-to-north transect: Terai, Hill, Mountain, and Trans-Himalayan regions. Nepal can be further divided into three major river systems, from west to east: Karnali, Narayani/Gandak, and Koshi, all tributaries of the River Ganges.

Table 4.5 Sector-Wise GHG Emissions in South Asia: Nepal

Emissions	In Percentages	In Tons
Transport sector	49	409,511.9
Residential sector	12	100,288.5
Industrial sector	25	208,934.5
Agricultural sector	8	66,859
Commercial sector	6	50,144
Total	**100**	**835,738.6**

Source: *From Lohani and Baral (2011).*

Climatically, Nepal has six zones from south to north subtropical (height <1000 m), tropical (1000–2000 m), temperate (2000–3000 m), subalpine (3000–4000 m), alpine (4000–5000 m), and Nival (>5000 m) region. The major climatic processes in Nepal have been dominated by monsoon. However, the impact of climate change on Nepal's water resources is unclear due to the uncertain behavior of glaciers. According to some estimates, roughly 20% of glacier mass could be lost with an increase of 1 °C in temperature. These projections, however, are still subject to debate. The recent trend shows that mean annual temperature has been inconsistent, so also the mean annual rainfall, which has been decreasing by −3.7 mm/month/decade, and cold days and cold nights have been reduced by 5 and 8 days, respectively. Winters are largely dry, while monsoon brings more than 1200 mm rain between June and September (SAARC, 2014).

The Nepal's National Adaptation Plan of Action (NAPA) prepared in 2010 aimed to achieve economic and social progress, while recognizing that climate will be uncertain and vulnerability will continually increase in this Himalayan kingdom (Dixit, 2011). As Climate "adapted" systems will help build social resilience, the ability to reduce vulnerability to disasters is related to the robustness of such adaptive systems (Moench and Dixit, 2004). A dearth in hydrometeorological information, coupled with lack of climate change monitoring station, it becomes difficult to adequately capture the temporal and spatial dynamics of precipitation in Nepal. Global circulation model (GCM) suggests increase in temperature in Nepal between 0.5 and 2.0 °C by 2030 and between 3.0 and 6.3 °C by 2090 (Chaudhary and Aryal, 2009).

The annual GHG emission by Nepal contributed by various sectors is shown in Table 4.5. The global climate scenario suggests that the impacts of climate change may be intense at high elevations and in regions with complex topography, as is the case in Nepal's mid-hills (IPCC, 2014; NCVST, 2009; Shrestha et al., 1999). This is likely to change river flows, which in turn will affect low flows, drought, flood, and sedimentation processes (Mirza and Dixit, 1997). Moreover, about 75% of the Nepal's population depends on agriculture that covers only 21% of the country's land area of which 65% area is only rain-fed. Hence bringing more area under irrigation will be a major job of NAPA. Financially insulating such a large population from climate impact will be a challenging task. The mitigation strategy must include speedy disbursement of loans, microcredit, crop insurance and arrange access to local markets. Seeing the climatic uncertainty, many wondered that whether it would be feasible to import food grains instead of producing crops an uncertain scenario.

The migration of people to India or to other countries reduces risk and in the short term can contribute financial resilience through remittances and reduced reliance on land-based livelihoods. However, migration also alters community relationships and local resource management dynamics. The long-term implications of such a strategy could weaken the social and economic health of Nepal as domestic skill and expertise would decline. At the same time, increased income from remittances can fuel an increase in consumer spending and creating a fragile domestic economy (NCVST, 2009; Verweij and Thompson, 2006).

To address the issues of climate change, Nepal initiated the national plan NAPA and LAPA (Local Adaptation Plan of Action), as part of its Climate Change Policy 2011 (Tiwari et al., 2014). Recently, the ministry of population and environment has communicated to the UNFCCC secretariat in February 2016 the INDC report with its 10-point program. This includes formulation of post-2020 NAPs to strengthen the Environmental-Friendly Local Governance (ELFG) framework in village development committees and municipalities to complement NAP, NAPA, LAPA policies. Simultaneously, Nepal will encourage scientific research to develop and implement adaptation strategies for climate change affected sectors to ascertain the actual loss and damage associated with climate change impact. Furthermore, Nepal plans to formulate the Low Carbon Economic Development Strategy, focusing on: energy, agriculture-livestock, forests, industry, human settlement-wastes, transport, and commercial sectors.

In addition, by 2050 Nepal plans to achieve 80% electrification through renewable energy sources, implementing National Rural and Renewable Energy Program (NRREP- under Alternative Energy Promotion Centre) reducing Nepal's dependency on biomass and making it more efficient. For example, Nepal intends to increase its mini and micro hydropower generation to 25 MW, solar Photo-Voltaic cells and pumping system to 600,000 numbers for home and another 1500 for offices. Also NRREP is to produce 4000 water mills, 475,000 cooking stoves, and 200 community biogas plant units,

Nepal plans to develop its hydropowered electrical rail network by 2040 to support mass transportation of goods and public commuting. Despite doing all these, Nepal has promised to maintain 40% of the total area of the country under forest cover and increase productivity and products from forests through its sustainable management. Emphasis will be equally given to enhance carbon sequestration and forest carbon storage and improve forest governance. Nepal also plans by 2025 to decrease the rate of air pollution through proper monitoring of sources of air pollutants like wastes, old and unmaintained vehicles, and industries (Fig. 1.4).

4.1.7 Pakistan

Pakistan has a profound blend of landscapes varying from plains to deserts, forests, hills, and plateaus, ranging from the coastal areas of the Arabian Sea in the south to the mountains of the Karakoram Range in the north (Fig. 1.4). Pakistan geologically overlaps both with India (Sind and Punjab) and Eurasian (rest all other provinces) tectonic plates. Pakistan is divided into three major geographic areas: the northern highlands, the Indus River plains and the Baluchistan plateau.

Pakistan lies in the temperate zone, and the climate varies from tropical in the south to temperate in the north, and is visited by monsoon. Rainfall varies from as little as less than 25.4 cm to over 380 cm in a year, in various parts of the country. These generalizations should not, however, obscure the distinct differences existing among particular locations. For example, the coastal area along the Arabian Sea is usually warm, whereas the frozen snow-covered ridges of the Karakoram Range and of other mountains of the far north are cold year round.

The recent trend shows that mean rainfall has decreased by 10%–15% since 1960 in southern Pakistan (it however increased in northern Pakistan during the same period), while hot days and hot nights have been increased by 20 days and 23 nights. The major climatic processes have been monsoon (heavy rain), El Nino (drier weather), Western depression (irregular rainfall), and SST in Arabian Sea regulating temperature (SAARC, 2014)

The important climate change threats to Pakistan are: (1) considerable increase in the frequency and intensity of extreme weather events, coupled with erratic monsoon rains causing frequent and intense floods and droughts, (2) projected recession of the Hindu Kush-Karakoram-Himalayan (HKH) glaciers due to global warming and carbon soot deposits from trans-boundary pollution sources, threatening water inflows into the Indus River System (IRS), (3) increased siltation of major dams caused by more frequent and intense floods, (4) rising temperatures resulting in enhanced heat and water-stressed conditions, particularly in arid and semiarid regions, leading to reduced agricultural productivity, (5) further decrease in the already scanty forest cover, from too rapid change in climatic conditions to allow natural migration of adversely affected plant species, (6) increased intrusion of saline water in the Indus delta, adversely affecting coastal agriculture, mangroves, and the breeding grounds of fish, (7) threat to coastal areas due to projected sea level rise and increased cyclonic activity due to higher sea surface temperatures, (8) increased stress between upper riparian and lower riparian regions in relation to sharing of water resources, (9) increased health risks and climate change induced migration.

Pakistan, among the South Asian countries, has the most comprehensive, well-laid out program, objectives, strategy, and detailed action plan toward mitigation and adaption with regard to climate change. The Pakistan Meteorological Department (PMD) shows that the mean surface air temperature in Pakistan has risen at the rate of 0.099 °C per decade from 1960 to 2010 resulting in total change of 0.47 °C, which is statistically significant at the 95% level of confidence. The warmest year in Pakistan recorded by the PMD was 2005 and the second warmest was 2007. Drastic rise in temperature in the last decade has been observed, which made it the warmest decade on record in Pakistan. Actually, Pakistan has moved from a water-affluent country to a water-stressed country. In 1947, per capita water availability was 5000 m^3, which has currently decreased to around 1000 m^3, and projected to decrease further to 800 m^3 by the year 2025.

The 93-page framework of implementation of National Climate Change Policy (NCCP, 2014–2030) in Pakistan prepared by its Climate Change Division in November 2013 was developed keeping in view the current and future anticipated climate change threats and adaptation issues in the country. The major survival concerns for Pakistan are particularly in relation to the country's water security, food security, and energy security. The NCCP has been designed to take action under four time-bound sectors: Priority Actions (PA): within 2 years, Short-term Actions (SA): within 5 years, Medium-term Actions (MA): within 10 years, and Long-term Actions (LA): within 20 years.

Pakistan's GHG emissions are, however, low compared to international standards. For example, in 2008 Pakistan's total GHG emissions were 310 million tons, which comprised carbon dioxide (CO_2) 54%, methane (CH_4) 36%, nitrous oxide (N_2O) 9%, carbon monoxide (CO) 0.7%, and nonmethane volatile organic compounds 0.3%. The majority of the CO_2 emission is accounted by power sector (32%), followed by industry (30%) and transport (22%) leaving other sectors to account for the remaining 16% SAARC (2016). The NCCP includes 737 specific action plans spread over 145 strategies covering 46 major objectives in various sectors to mitigate climate change and adapt new initiatives to reduce GHGs in atmosphere. The total GHG emission by Pakistan is given in Figs. 2.1–2.6.

Pakistan published her INDC in 2016, which presents the overall GHG emission profile and future emission projections of the country, by considering changes in the demographic dynamics and emerging energy requirements (Pakistan-INDC, 2016). It also describes mitigation and adaptation measures, vis-à-vis challenges likely to be confronted in coming years. Low abatement cost coupled with prospects for climate-resilient investment in energy, industry, and flood infrastructure including water reservoirs and water channels, qualify Pakistan as one of the promising carbon investment markets in the world. According to a national study, Pakistan's annual adaptation necessity is between 7 and 14 billion US$.

4.1.8 Sri Lanka

Sri Lanka has a total area of 65,610 km^2, with 98% land surface. Its coastline is 1340 km long, and total population has been 20,277,597 (as on 2012, Fig. 1.4). Sri Lanka is connected to India by Adam's Bridge in the north, a causeway now submerged with shoals of limestone. The natural resources found are graphite, gems, phosphate, clay, and hydropower.

Sri Lanka has a tropical climate (humid and warm) and displays the following climatic periods: (1) SW monsoon period (May–September): 55% of the annual precipitation and often exceeds 3000 mm rain, (2) intermonsoon period (October–November) causing about 500 mm of rain, (3) NE monsoon period (December–February) causing 200–1200 mm of rain, and then (4) intermonsoon period between March and April. While El Niño causes drier weather, La Niña causes heavy rainfall, and Intertropical Convergence Zone (ITCZ) also drives rainfall. The temperature varies from 16°C in the central highlands to 32°C in the coastal regions (average 28–30°C). The recent trend shows that mean annual temperature has been increased by 0.16°C per decade since 1960, while the mean annual rainfall has been reduced by 144 mm since 1960 compared to the period between 1931 and 1960. Among the environmental issues, deforestation, soil erosion, urbanization, coastal degradation, and water and air pollution are of much concern to Sri Lanka (SAARC, 2014).

Being a developing island nation and characterized by tropical climate patterns, Sri Lanka is highly vulnerable to climate change impacts. Extreme weather events such as high-intensity rainfall followed by flash floods and landslides, and extended dry periods resulting in water scarcity are now becoming common issues in Sri Lanka. The GHG emission from Sri Lanka is depicted in Table 4.6 and Figs. 2.1–2.6. According to the National Climate Change Policy of Sri Lanka (NCCP), any adverse changes in already volatile weather patterns are likely to have a severe impact on the socioeconomic activities in the country. Therefore, urgent action is necessary to build resilience to lessen the undesirable impacts of climate change. Accordingly, NCCP of Sri Lanka aims to: (1) sensitize and make aware the communities periodically on the country's vulnerability to climate change, (2) take adaptive measures to avoid/minimize adverse impacts of climate change on livelihoods and ecosystems, (3) mitigate GHG emissions in the path of

Table 4.6 Sector-Wise GHG Emissions in South Asia: Sri Lanka

Different Sectors	Percent of Total Emissions (%)
Energy	65
Agriculture	23
Waste	10
Industry	2

Source: *Modified from Ranasinghe (2010).*

sustainable development, (4) promote sustainable consumption and production, and integrate climate change issues in the national development process, and (5) develop the country's capacity to address the impacts of climate change effectively and efficiently.

The NCCP wants to achieve these objectives through several intense measures, such as: (1) vulnerability (coastal erosion, health impairment, and disaster management), (2) adaptation (on food production and security, rainfall shortage, watershed and wetland management, human settlement and land use planning, infrastructure design and development), and (3) mitigation (create and use clean and renewable energy, and encourage low emission transportation, ensure low emission from industries through green audit, integrate waste management system, and promote appropriate innovative technology for agriculture and livestock management). The NCCP further attempts to encourage sustainable consumption, production and lifestyle, knowledge management, and institutional coordination.

Sri Lanka has empowered the Ministry of Mahaweli Development and Environment to lead and to also act as the National Designated Entity (NDE) for the United Nations Framework Convention for Climate Change (UNFCCC). Two major milestones of this national initiative are the National Climate Change Adaptation Strategy for Sri Lanka prepared in 2010 and the NCCP formulated in 2012. Sri Lanka has already started the National Adaptation Plan for Climate Change Impacts (NAP). In addition, Sri Lanka has initiated several actionable programs, such as Harita Lanka (greening the Island), comprehensive disaster management, combating degradation of lands, coastal zone management, country's national physical plan, water development plan and draft agricultural policy.

4.2 SHORTCOMINGS

It appears that almost all the countries in South Asia have woken up adequately to the profound threat of climate change. The adaptation and mitigation plans and programs chalked out by each of these governments appear to be well thought of. Various impacts owing to climate change on South Asia are summarized in Table 4.1.

Better-performed missions in India have been the National Solar Mission (target generation of >3000 MW) and the National Mission for enhanced energy efficiency (saved 11000 MW of power in the last 5 years). However, slow take-off of several missions, namely, green India (aforestation suffering due to lack of funds), agricultural development, urban growth (building and vehicle codes), Himalayan ecosystem, and acquisition of strategic knowledge has made the struggle for mitigation difficult. The power generation status and commercial energy resources in South Asia are assessed (Table 4.7).

The agriculture is the mainstay of South Asian economy. It is, hence, natural to note that any excess in parameters such as carbon dioxide, other greenhouse gases, temperature, precipitation, and glacial run-off would impact agricultural production and practices. Conversely, agriculture releases greenhouse gases (methane) and alter the coverage of land through deforestation/ desertification. All these could change the Earth's ability to absorb or reflect heat and light. Agriculture and livestock-related activities contribute in increasing greenhouse gas concentration in atmosphere through (1) CO_2 release from deforestation, (2) CH_4 release from paddy cultivation and fermentation in cattle, (3) N_2O releases from fertilizer applications, and (4) nitrogen leaching from soil drainage.

The IPCC (2007, 2014) predicted a decrease in agricultural production in South Asia by 30% by the middle of this century. It seems that an increase by even a degree in temperature could cause a 20% reduction in yields in South Asia, as rice becomes sterile if exposed to temperatures above 35°C even for an hour during the flowering period. Deforestation, variation in agricultural land cover, burning of vegetation and consequent albedo effect contribute largely to GHG emissions and soot formation. The key indicators of growth of South Asia as projected till 2050 have been estimated (Table 4.8)

The change in agricultural land in South Asia over the decades shows that except Bangladesh, all the other seven countries in South Asia have increased agricultural land cover in 2005 compared to that in 1965. In fact, Maldives has doubled the area covered by agriculture, while Bhutan came close to second. Sharp decline in land cover used for agriculture in Bangladesh is certainly of concern.

However, besides crop-climate syndrome, the psychology of the farmers as well as of government policies could play major role in ensuring food security vis-à-vis GHG emissions. An overall lack of interest toward agriculture has been noticed in almost all the South Asian countries. For example, while the total population in South Asia had increased to just under 1700 million in 2010, the agricultural population had increased merely from 600 in the year 1980 to less than 750 million in 2010 (75 crores) suggesting that only 30% of the total population was involved in agriculture in 2010 (FAO, 2015). A comparison of rural, urban, and agricultural population vis-à-vis agricultural land cover in 1960 with that in 2010 shows consistent decrease in agricultural land. Con-

Table 4.7 Power Generation Mix and Commercial Energy Resources (#) in South Asia

	Afghanistan	Bangladesh	Bhutan	India	Maldives	Nepal	Pakistan	Sri Lanka
Coal	3,582,040	1,225	0	869,181	0	0	142	3,506
Oil	na	6,692	0	23,169	330	10	550,607	4,439
Gas	293	404,083	0	65,102	0	0	96.71	0
Biofuels	0	0	0	21,809	0	0	0	41
Waste	0	0	0	1,338	0	0	0	0
Nuclear	0	0	0	34,228	0	0	0	0
Hydro	125,000	894	7,164	141,637	0	3,636	225,000	4,552
Solar PV	0	145	0	3,433	0	0	0	20
Wind	0	4	0	33,583	0	0	0	272
Total	3,707,333	413,043	7,164	1,193,480	330	3,646	2,800,845	12,830
Imports		0	0	5,598	0	1,072	0	0
Exports		0	5,147	5	0	3	0	0

#, Original Mtoe values converted to Gwh; PV, photovoltaic.
All values in gigawatt-hour (GWh).
Wijayatunga and Simbalapitiya (2016) and Wijayatunga and Fernando (2013)

Table 4.8 Projected Key Indicators of South Asia

Indicators	Years	2050 Projection
Population (millions)	1,520 (2007)	2,242
Annual population growth rate (%)	1.5 (2007)	0.53
Population below US$1.25 a day (millions)	596 (2005)	14.1
Undernourished population (%)	21.8 (2007)	4.2
Irrigated area (million ha)	104.3 (2000)	135.2
Total rain-fed area	98 (2000)	110
Fertilizer consumption (kg/ha)	210 (2007)	256
Agriculture growth rate (%)	2.4 (2007)	1.3
Crop production growth rate (%)	2.1 (2007)	0.9
Cereal production growth rate (%)	1.9 (1997)	—
Livestock production growth (%)	3.2 (2007)	2.2
Total water consumption in agriculture sector (km^3)	1,479 (2000)	1,922

Source: *Modified from De Fraiture and Wichelns (2007); FAO, 2012. The state of food and agriculture Food and Agricultural Organization of the United Nations, Rome, pp. 1–182; Lal (2007); Rasul (2014).*

sequently, the generation of methane from paddy cultivation since 2008 has come down drastically. What in 1990 was only 129,000 Gg, the methane generation increased to 135,000 Gg in 1998, went down thereafter to 125,000 Gg in 2001, to increase again to 135,000 Gg in 2008, only to drastically fall to 120,000 Gg in 2010 (Ritchie and Roser, 2017).

In summary, there have been substantial shortcomings between framing of policies and the exact implementation of such policies on the ground. The 2004 Tsunami in the Indian Ocean, devastating floods of India (Uttarakhand, Kashmir, Mumbai), Pakistan, and earthquakes of Nepal, north India, and Pakistan are some of the major natural calamities that exposed the inadequate, incoherent, ill-coordinated government machinery that were responsible to alleviate the impact of natural vagaries. Although a clear increase in number of tragedies befalling South Asia over the decades have been recorded, the loss of human lives has been reduced.

The reasons for the gap between policy and implementation are manifold, such as: (1) the framed policies are too formal, (2) the uncertainty in IPCC data and model on global warming, (3) lack of trained human power (capacity building), and (4) lack of funds to implement perceived climate policies. The ground veracity has been quite different than what is being apparent, in terms of innovation in mitigation, and coordination in reaching relief quickly to the victims of natural disasters. Moreover, many government departments in India and in South Asia are at loggerheads on the long-term low carbon growth policy. The slothful implementation of programs that help people adapt to climate change impacts, complacent attitude, and lack of coordination between the federal and the provincial governments have made the matter worse.

The global warming politics is extremely polarized. The link between every stage of economic progress (production, transportation, storage, delivery, and disposal of any item) in a country and its GHG emission is intimate. And in the absence of advanced carbon-neutral technology, particularly in South Asia, the regions dependence on fossil fuel will continue. This would sum up to summarily increase in CO_2 content in atmosphere. However, the per-capita GHG emission by South Asia will remain low than many OECD (*Organization of Economic Cooperation and Development*) countries.

In addition, the developing countries, including South Asia, see climate change as a hindrance to their path of economic development. These countries blame the developed world for having created the global warming crisis by emitting increased GHG during the industrial revolution in 19th and 20th Centuries, and suggest that developed countries must now pay to mitigate and repair the damage. As global negotiation on climate change aims for consensus, it would remain difficult to enforce any decision or agreement. The print and electronic media make things worse at times by sensationalizing reports of climate change.

From the very high coal consumption in India to intensive deforestation in Pakistan (the highest rate in Asia) to the rampant use of heavily polluting biomass fuels across the entire South Asia are alarming. Although South Asia's carbon footprint is lighter relative to other regions, it is high time this region moves to obtain energy from solar radiation. If not, South Asia's carbon footprint could more closely approximate that of the West.

Nevertheless, there are some turnaround stories too. The construction of the world's largest single solar photovoltaic power plant in Bahawalpur in Pakistan's Punjab province with Chinese collaboration shows Pakistan's commitment to mitigate climate change. This 100 MW power plant has come up over 229 ha at a cost of US$215 million. Similarly, India proposed the International Solar Alliance (ISA), an intergovernmental coalition of more than 125 sunshine (*Suryaputra*) countries, which lie between the Tropic of Cancer and the Tropic of Capricorn. The primary objective of the ISA is to work for efficient exploitation of solar energy to reduce dependence on fossil fuels. Later, the Framework Agreement of the International Solar Alliance (FAISA) opened for signatures in Marrakech, Morocco in November 2016, and 121 countries have already joined the ISA. India, for example, indeed plans to get approximately 40 percent of installed electric power capacity from non-fossil fuels by 2030, and firmly pushing forward Global Solar Alliance.

While advocating Yoga as an active mechanism to reduce carbon footprint, India did not agree to the developed world's concept of "climate neutrality," which it believes would give developed economies a license to keep polluting. However, in complete contrast to this apprehension, India laid out in its INDC specific plans to attain a 33%–35% reduction in emission intensities per unit. Emission intensity is the ratio of carbon dioxide to a measure of economic output (it is, however, not a measure of absolute emissions).

In fact, implementation of the Paris Agreement in South Asia would require high-end technologies, transformative policies and new measures to stabilize warming in the range of 1.5–2°C (Ravindranath et al., 2017). Consequently, the South Asian countries, together under the SAARC umbrella or independently, may bring about change in strategy for mitigation and adaptation, say, in agriculture, urban planning, ecosystem management, and industrialization. Similarly, necessary modifications in regulations to finance and manage the transfer of wealth will be of immense importance.

CHAPTER 5

Threat To Opportunity

It is said that civilization which can take care of the essential needs of its people will only prosper in future. The current as well as future needs of people will be mainly energy-centered. History teaches us how civilizations suffered when energy, food, and water, the three most essential items, were not made available in ample supply to the people. The present civilization will also suffer hugely, closing to extinct, if these three resources could not be made available adequately. Hence, the focus of this chapter is to examine whether human beings can turn the threat emanating from climate change to opportunity, ensuring in the process inclusive growth, infrastructure expansion, and human development. Later the advancement made so far on technology front to generate inexpensive, sustainable, environment-friendly, job-intensive renewable energy industry will be touched upon. Also the possibility of newer technologies of climate engineering is examined.

5.1 OPPORTUNITIES A-GALORE

The past has shown us in innumerable times that new ideas and innovations emerge only when the people are hard pressed and forced into corner. In this regard, two initiatives will be essential—developing climate change mitigation technologies (CCMT), and changing the mindset of people (including doing similar/same things differently). Although greenhouse gas (GHG) emissions could have been partially arrested with legal and economic initiatives of the governments of the world, climate change is a slow, subtle, but continuous process, which seems unstoppable. Hence, mitigation and to a larger extent, adaptation, are probably the real answers to alleviate the damages that are expected from the change in climate.

However, the wide-gaps between the mitigation strategies developed by the respective South Asian nations and their implementation on the ground (Section 4.2) make it clear that the growth strategy that could be prescribed

Climate Change. http://dx.doi.org/10.1016/B978-0-12-812164-1.00005-0

for the economically advanced countries may not be necessarily applicable in South Asia. Hence, by keeping an eye on the resources available in South Asia, in addition to those that these nations could access from the global market, it is believed that the threats of climate change to South Asia may probably be turned into opportunities, in the following manner—(1) invest optimally in innovating newer climate technology gadgets, which in turn will create new types of jobs, (2) jobs and services consequently will enhance social protection, alleviate poverty and inclusive economic growth in South Asian society, and (3) such investment would require collaboration and cooperation among nations, which would pave way to remove uncertainty from market and strengthen economic and political stability in the entire region (Table 5.1).

The companies and investors that seize these opportunities will be best positioned to thrive in a resource-constrained economy. In addition, such an initiative would simultaneously keep environment healthy, create new-types of employment, and would make individuals wealthy. Toward this, generating electricity efficiently and in large quantity with minimum emission of CO_2 has been the best initiative.

It has been heartening to note that private sectors are taking interest in combating climate change impact. Many of these private companies, having financial clout and affinity for innovation, must play a leading role in the struggle for a greener future (Tsitsiragos, 2016). These private companies assured the world fora (COP 21-23) of their commitment to address climate change, and made pledges to decrease carbon footprint, buy more renewable energy and engage in sustainable resource management. What the private sectors want in exchange from the governments is a stable, time-tested and sustainable public policy and long-term regulations including a price on carbon.

The rapid improvement in technology is making manufacturing more capital and skill intensive. In addition to respecting domestic demands and priorities, such technology improvement must address regional and global requirements (Rodrik, 2013). However, without appropriate care, such development can go for a spin denying in the process the deeper political and cultural roots of the society. Hence, a soft adaptation could go a long way to address this issue (Pelling, 2011).

Hence, like nature, GWCC (global warming and climate change) could also be seen as best transformation-vehicle and also opportunity to bring in socialism and may be responsible for distributing wealth among the all sections of the society, legally. The battle is difficult, and challenging, but these threats could be turned into opportunities, if sincere governance powered by greedless ethical culture and values is designed.

Table 5.1 Climate Change: Threat To Opportunity Scenario in South Asia

What is needed in this world now is to Integrate all aspects of nature for Harmonious Growth of Civilization		• PHYSICAL RESOURCES (Infrastructure, Energy, Mining, Tourism, Trade and Commerce) • NATURAL RESOURCES (Land, Ocean, River, Springs, Mountain, Desert, Soil, Wetlands) • HUMAN RESOURCES (People—Education, Health, Food, House, Potable Water)
Threats	**Solutions**	**Opportunities (Each of these activities would create jobs): Activities, Technology use and Creation of Job and Facility**
Shortage of water to Irrigate	Pond sand filter	• Ensure clean water supply for domestic use in times of drought; • Provide easy access to water source, saving time.
	Rainwater harvesting	• Meet basic needs for clean drinking water; • Stores clean water for use during times of drought.
	Pond water harvesting using solar PV pump	• Pump water to crop lands from ponds during times of drought; • Reduce use of fossil fuel dependant pumps.
Lack of Education	Establish Information and technology exchange centre	• Share, transfer and spread innovative ideas and technology related to climate change adaptation; • Demonstrate and train farmers on improved seed varieties; • Raise awareness on climate change issues within the broader community.
	Education at early stage	• Educate children about climate change and disasters.
	Mobile Library	• Raise awareness and educate hard-to-reach communities through fun
	Eco-clubs and weather stations	• Increase the level of understanding of the hydro-meteorological sector and to disseminate information on climate change in the local community.
Lack of Skill	Floating gardens	• Assist farmers grown crops in times of flood and inundation.
	Crab cultivation	• Move communities from shrimp farming to crab cultivation and fattening.
	Cluster housing	• Raise the level of a group of houses to reduce risk of flooding; • Encourage communities to engage in a range of different productive economic activities to increase their resilience to disasters.
	Farmer field schools	• Provide a forum for learning, sharing and training in local communities.
	Caged fish cultivation	• Provide a source of protein and income for farmers without access to ponds, and during times of flood.

(*Continued*)

Table 5.1 Climate Change: Threat To Opportunity Scenario in South Asia (*cont.*)

Lack of Water and Health	Flood proof tube wells	• Ensure year-round access to safe drinking water
	Flood proof sanitary toilets	• Ensure sewerage is contained within the latrine during times of flood
Floods and Soil erosion	Construction of embankments for flood protection,	• Protect agricultural lands from inundation
	Community flood shelter	• Provide temporary shelter from floods for community members and their livestock
	Plantations to control river erosion	• Reduce flood damage and provide livelihood support for communities during flooding season
	Tidal River Management	• Raise waterlogged areas using sediments brought in by tidal flows for agriculture
	Duck rearing	• Provide sources of protein and income during times of flood
	GLOF warning system	• Warn people living in downstream areas of GLOF events
Agriculture, Seeds, Crop, Soil	Community based seed conservation	• Keep seeds safe from flood water • Management of Crop, Soil, Land, Irrigation, Pest, Food, Storage • Skill & Capacity building of farmers to newer techniques
	Minimum tillage and crop residue management	• Crop diversification, flood and drought resistant seeds • Increase soil organic content, Sprinkler and drip irrigation
	Sloping Agricultural Land Technology (SALT)	• Maintain soil health and control insect pest populations
Smog inside home	Portable and more efficient cooking stoves	• Facilitate cooking during times of flooding • Reduce the incidence of smoke that has negative health effects for women • Reduce the amount of fuel required, lessening the impact on environment
Cross-sectoral	Research and Development	• Climate Modeling and Scientific research, Technology innovation • Early warning system and Disaster communication • Hazard insurance for crop against flood, drought, fire and pest • Financing adaptation, Planning & Strategy development

Source: *Adapted from Defra (2009) and Sterrett, C., 2011. Review of Climate Change Adaptation Practices in South Asia, with the permission of Oxfam, Oxfam House, John Smith Drive, Cowley, Oxford OX4 2JY, UK. www.oxfam.org.uk. Oxfam does not necessarily endorse any text or activities that accompany the materials, nor has it approved the adapted text*

5.2 INNOVATIONS IN CLIMATE TECHNOLOGY

The scale and urgency of the problem of GWCC are far more serious than even many climate policy advocates acknowledge. It is further insisted that solution to the future energy needs of South Asia (and also of the world) lie not in coal, oil, and nuclear power (all these must be phased out as quickly as possible), but in greater use of renewable energy sources for both heat and electricity. Renewable energy technologies are developing and maturing quickly. The clean technology fund of the Asian Development Bank is investing about US$880 million in renewable energy projects in Asia, and another US$255 dollar in sustainable transport projects. In terms of region, Southeast Asia will receive 52% of ADB funds, followed by South Asia (44%) and Central and West Asia (4%; ADB, 2014). South Asia must make maximum use of such largesse. Alternate energy from nonhydrocarbon resources is discussed below to show the spectrum of new opportunities (also see Tables 4.2 and 5.1).

5.2.1 Renewable Energy

Compared to energy received from coal, oil, and wood, the renewable energies are largely clean in nature, and cause minimal pollution and release of GHG (Table 5.2). Renewable sources contributed about 19.2% of global energy consumption in 2014, which has increased every year since then. Of this, about 9% came from traditional biomass, and about 2% each from modern biomass, geothermal, solar, hydroelectricity, and wind source (IEA-REN, 2016).

The world invested about US$286 billion in 2015 toward the production of renewable energy from wind, hydro, solar and biofuels, with the USA and China spending more than others. The patent data with regard to climate change mitigation technologies (CCMT) derived from *European Classification System-ECLA, European Patent Office-EPO*, and *Worldwide Database of Patent documents-PATSTAT*) shows that about 15% of patents are filed to fabricate new generations of cost-effective solar cells and hybrid solar photovoltaic technology. Similarly, about 13% patents are filed to manufacture instrument to reduce carbon emission from factories, 11% patents on manufacturing fuel cells, about 9% on to ensure energy efficiency in building, and another 12%–13% each to produce emission abatement and fuel efficiency, wind turbines and battery, hydrogen technology, and fuel cells. Each of these patents comes with countless opportunities down the system.

It is estimated that about 7.7 million jobs are estimated to be available worldwide in the renewable energy industries, with photo-voltaic (PV) cell manufacturers being the largest employer (Shukla et al., 2017). In fact, more than half of the new electricity capacity installed after 2015 is renewable (The Guardian, 2016). The advantages of renewable energy over the existing conventional fossil fuels are listed in Table 5.2. The governments of South Asian countries have initiated renewable energy policies to encourage industries and individuals to employ renewable energy powered systems.

5.2.1.1 Solar Radiation

It is estimated that the annual solar radiation reaching the Earth's surface could equal 10,000 times the world's yearly energy needs (Goswami and Besarati, 2013). In many cases solar electricity is already cost competitive. The cost of manufacturing solar cells and modules and other components has been falling steadily. Actually, the price of PV systems has fallen by an average of 5% per annum over the last 20 years (Greenpeace, 2007). Solar energy could be put into use in several areas—thermal collectors to warm air, water, and PV collector to generate electricity directly from sunlight. Desalination, powered by solar energy, in large scale could solve the impending potable water problem of the world.

Table 5.2 Advantages and Disadvantages of Renewable Energy

	Advantages of RE	Disadvantages of RE
1	Source is renewable, infinite, sustainable	Initial investment for RE plant is too high
2	Cleanest energy, with no or little GHG emission No impact on environmenthealth. No production of waste	Availability of RE heavily depends on unpredictable weather and supply may break when one needs it most
3	Dependence on fossil fuel has compromised the national Security of many South Asian countries It brought in at times political instabilities, trade disputes, wars, and high prices	It is difficult to generate RE in large quantity from a small-area plant like that of thermal/nuclear power
4	Cheap, competitive, low maintenance, and would create millions of jobs	RE requires large stretch of land to produce large quantity of electricity. Naturally the land is in prime demand. And with consistent increase in population, the demand for land to carry out agriculture will only increase making RE unviable
5	RE will offer a steady energy price as it would have no raw material cost	
6	Solar cells could be placed on building, trains, moving vehicles, as pavement roof, over large water bodies	

RE, *Renewable energy.*
Source: *(IEA-REN, 2016; UNFCCC, 2017).*

How South Asia is blessed with regard to solar radiation could be gauged from the fact that the skies of Afghanistan remains clear with sunshine for about 300 days a year, and it is estimated that solar radiation averages about 6.5 KW-hours per square meter per day in Afghanistan (Shukla et al., 2017). Consequently, the potential for solar energy development is so huge that such solar power could be distributed throughout Afghanistan and even to other areas in South Asia through solar-grid (Meisen and Azizy, 2008). Under the Renewable Purchase Obligation, India is set to produce 17% of her electricity from solar radiation by 2022, including adding 48 GW (gigawatts) of electricity through new solar capacity by 2019 (IEA-REN 2016). The formation of International Solar Alliance (ISA) in this regard is a huge step and will be discussed later.

5.2.1.2 Ocean Energy

The main forms of ocean energy are wave energy, thermal energy, tidal energy, and ocean currents. Electricity could be generated from the movement of waves, tides, currents, and in temperature and salinity gradients encountered down the water column. Tidal energy is obtained from the rise and fall of tides. As the tide fluctuates, a massive amount of water moves toward and then away from shore. These strong waves could turn giant turbines placed in the path of the moving water. These spinning turbines are connected to generators that create electricity (WEC, 2013). In case of tidal barrages, the energy generation takes advantage of the difference in height between low and high tides.

Tides are more predictable than wind energy and solar power. Moreover, tidal energy has a high energy density, implying that the tides store a larger amount of energy than most other forms of renewable energy. Ocean energy technologies—both wave and tidal energy are—being developed actively in South Asia. India has a potential to generate little more than 8,000 MW of tidal energy. This includes about 7,000 MW in the Gulf of Cambay in Gujarat, 1,200 MW in the Gulf of Kutch and 100 MW in the Gangetic delta in the Sundarbans region of West Bengal. The estimated cost for generation of electricity from tides would be about INR 1/- per unit with an investment of 1,460 crores to generate electricity. As a part of this campaign, in 2014, it was proposed to install and develop 50–200 MW tidal stream based power plant at Gulf of Cambay, Gujarat. However, the high initial investment cost and objections from environmentalists are holding back the entire tidal energy technology from being increasingly implemented (Srivastava, 2016). The Maldives could examine the challenging potential to use large swells to generate electricity.

South Asia being a tropical region experienced a constant difference also in temperature between surface water and the deep ocean. This gradient can be used to generate power and fresh water simultaneously (Yadav, 2015). SAARC may build floating offshore ocean thermal energy conversion (OTEC) plant of power generation of larger capacity to serve coastal mainland and Islands.

5.2.1.3 Wind Power

The generation of electricity from wind is competitive, and least polluting. Wind provided nearly 74,000 MW of power in the United States in 2015, a five-fold increase since 2008. It seems that if the United States obtains 21% of its electricity from wind power by 2020, the measure will reduce global warming emissions the equivalent of taking 71 million cars off the road or planting 104 million acres of trees. Moreover, it is estimated that every MW-hour produced by wind energy avoids an average of 1,220 pounds of CO_2 emissions (Audubon, 2016).

South Asia has many deserts, particularly in western India, Pakistan and Afghanistan, wherein wind power could be harnessed effortlessly (Meisen and Azizy, 2008). In hinterland areas of South Asia, campaign is on to install wind turbines on the top of mountains and hills. Moreover, South Asia has a long coastline, which has been the ideal site for installing wind power generating facility (GWEC, 2012). For the proper utilization of wind energy, South Asia needs comprehensive wind mapping to determine its wind energy potential and locate ideal sites for large-scale wind farms. India has been much ahead compared to other South Asian countries in utilizing wind power and all can benefit immensely by cooperating and sharing Indian technological know how and experience in harvesting of wind resources (Audubon, 2016).

5.2.1.4 Hydropower

The Earth's water cycle could also be used to generate electricity. South Asia, being rich in number of rivers, has been utilizing this resource quite heavily. For example, hydropower contributed about 22.5%, that is, 41,632 MW of the total electricity generation in India in 2011 (EAI, 2015), and an additional facility to generate 13,000 MW is under construction. India is endowed with rich hydropower potential to the tune of 148 GW (which would be able to meet a demand of 84 GW even at 60% load factor) which makes it one of the most important potential sources to meet the energy needs of the country.

Currently, over 93% of the total hydropower potential in the Brahmaputra River Basin is not tapped (Sharma, 2015). Similarly, only about 125 sites have been identified in Afghanistan for micro hydro resources, with the potential to generate about 100 MW of electricity (Meisen and Azizy, 2008). According to world energy resources (WER, 2016) India has 540,000 GWh/y undeveloped hydropower potential, whereas Nepal and Pakistan had 205,777 and 172,820 GWh/y respectively. As far as of total potential hydropower is concerned India has 660,000 GWh/y, while Nepal and Pakistan had 209,338 and 204,000 GWh/y respectively. Therefore, hydropower plants, it seems, are in a way profitable both environmentally and economically (Bhoi and Ali, 2014).

5.2.1.5 Bioenergy

Bioenergy can be obtained from biological and renewable sources. And, South Asia offers a conducive environment for accelerating the use of bioenergy technologies. Nearly 25% of its primary energy comes from biomass resources, and close to 70% of rural population depend on biomass to meet their daily energy needs. Bioenergy is of two types—traditional and modern. The traditional type appears in solid form that includes fuel-wood, charcoal, wood pellets, animal-dung, briquettes and so forth. The modern bioenergy, in contrast, comes in liquid form that includes bioethanol and biodiesel (Gumartini, 2009).

(a) *Traditional Bioenergy*: The burning of biomass is still prevalent widely in South Asia for energy needs, despite the fact that biomass burning for cooking causes severe indoor air pollution impacting health, particularly of women and children. Hence, any critical technological improvement in the form of a cheap efficient oven that involves low-cost biomass burning with the least GHG emission could bring about a socioeconomic-health revolution in South Asia. The children, who were supposed to study, are being engaged in collecting wood-biomass, refusing in the process long-term HR development in the country. Although lately, respective governments in South Asia are gearing up to supply LPG stoves to rural households for cooking. The unsustainable use of forest biomass by industries and increasing demand for land has led to deforestation and change in land use that contribute about 40% of carbon emissions from anthropogenic sources (WRI, 1996).

(b) *Modern Bioenergy*: The modern type of bioenergy is produced through anaerobic digestion—a type of biological process—and is also known as biofuel/biodiesel. This ester-based fuel-oxygenate is derived from renewable bioresources such as jatropha, soybean, mustard, rapeseed, peanuts, palm oil, other vegetable oils, and animal waste like beef tallow (Raju, 2006). Biofuel takes a short period of time (days, weeks, or months) to prepare, compared to millions of years taken by fossil fuel. Biofuels can also be made through chemical reactions, carried out in a laboratory or industrial setting to use organic matter (i.e., biomass) to make fuel. Biofuel is cheap, environment friendly, and does not necessarily depend on vagaries of nature (Raju, 2006). Countries spending huge sums on oil imports would naturally be interested to introduce biofuels aggressively by cutting down the cost of production.

> In South Asia, India is promoting bioethanol and biodiesel through phased mandates, fixed prices, and tax incentives, with production picking up since 2006. This includes production of ethanol from sweet sorghum, sugar beet, cassava, and tapioca, and production of biodiesel from nonedible seed bearing trees/shrubs like jatropha. To avoid any conflict between food and biofuel (as they compete for the same land to grow), planting biofuel crops only on wastelands throughout the country is recommended. In addition, integrating such a production with rural development programs is being worked out (Institute for Global Environmental Strategies White Paper), to encourage further research and development on cultivation, processing and production of biofuels. Moreover a blending mandate of 20% ethanol and biodiesel by 2017 is being explored.

Besides India and Pakistan, Bangladesh is also gearing up and has allowed private companies to explore manufacturing fuel alcohol in the country. An investment of US$4.5 million will be made for a 12,000 L/day ethanol plant in Bangladesh, which would use molasses as a feedstock. This modern group of bioenergy uses highly efficient combustion technology under tight regulations on emissions (Lali, 2016).

5.2.1.6 Geothermal

Geothermal energy comes from the natural heat of the Earth primarily due to the decay of the naturally radioactive isotopes of uranium, thorium, and potassium. The heat flow on Earth's surface averages 82 mW/m^2 which amounts to a total heat of about 42 million megawatts. The total heat content of the Earth is of the order of 12.6×10^{24} MJ (1 MJ = 1 mega-joule = 1 million-joule = 277.777778 Watt-hours), and that of the crust, the order is of 5.4×10^{21} MJ (Dickson and Fanelli, 2004), while the Earth could use only 7.1×10^{13} MJ to generate electricity in 2007 (IEA, 2016). It means, the thermal energy potential of the Earth is immense, but only a fraction of it is being utilized.

The electricity generated from geothermal system incurs a low production cost with controlled GHG emission. The countries across the world are making big stride in the geothermal sector. In South Asia, the hilly areas of Pakistan, Afghanistan, Nepal and Bhutan, India, and states in plains- like Andhra Pradesh and Gujarat in India are coming forward to develop electricity from this resource in a big way.

Although India is considered to be in low geothermal potential region with low/medium heat enthalpy, the estimated potential for geothermal energy in India is about 10,600 MW. Such potential areas are spread over 340 hot springs. The low average cost of production will be about 30 crores per MW. With such low generating cost, 1 MW geothermal power plant annually can generate about 8.3 MU (million units) per MW, compared to solar 1.6 MU per MW, wind 1.9 MU per MW, and hydro at 3.9 MU per MW. In addition, the geothermal waters (hot-springs) from Bihar, Jharkhand, and West Bengal reveal that thermal gases in these sites contain highly enriched helium (1%–3%v/v; Chandrasekharam and Chandrasekhar, 2015). Commercial extraction of this helium is being examined (Singh et al., 2016).

Besides India, Afghanistan and Pakistan have high potentials to generate electricity from geothermal source. At present, both abundant high-temperature water or steam, and adequate, efficient and durable technology to harness power from such resources are available in Afghanistan. Also, the underneath tectonic structure of Afghanistan facilitates the formation, circulation and accumulation of geothermal resource.

5.2.1.7 Hydrogen

Hydrogen is the most basic and omnipresent element in the universe, which if used properly, has the potential to re-energize the socioeconomic progress. As it is the fundamental stuff of stars, when burned, hydrogen produces only heat and pure water, with almost no carbon emission. Hence, hydrogen is the clean and sustainable alternative to fossil fuels. It also reduces deforestation for firewood and charcoal. Around half of worldwide hydrogen production currently comes from methane (49%) of natural gas. The rest comes from hydrocarbon by-products of chemical and processing industries (28%), nearly 18% from coal gasification and only around 4% from water electrolysis.

Hydrogen gas produced by electrolysis of water (an electric current flowing through water) could be used for household cooking. Because it does not emit CO_2, hydrogen gas could bring an important improvement in health and quality of life in South Asian households (Topriska, 2016). Transporting hydrogen under high pressure is somewhat dangerous, but it can safely be done through metal hydride storage, which, however, is very expensive and needs newer research to develop an alternate mechanism of transport.

5.2.2 Global State of Affairs

Interestingly, the global energy-related carbon dioxide emissions from fossil fuels and industry were found nearly flat since 2014. This is due largely to the decline in coal use worldwide but also due to improvements in energy efficiency and to increasing use of renewable energy. As of 2015, renewable energy provided an estimated 19.3% of global final energy consumption. The renewable energy sector employed 9.8 million people in 2016, an increase of 1.1% over 2015. By technology, solar PV and biofuels provided the largest number of jobs. Employment tilted further toward Asia, which accounted for 62% of all renewable energy jobs.

The overall energy of renewable energy resources in South Asian countries indicates that the region has huge hydro-, wind-, and solar-power potential (Table 4.2; Iftikhar et al., 2015; Shukla et al., 2017). In fact, Nepal alone has a hydropower potential of 83,000 MW, and even if energy demand increases at a rate of 10%, domestic demand will reach only 3500 MW by 2025. This presents a profitable opportunity for Nepal for energy trade that will also help in enhancing the energy security in the South Asian region as a whole (Rahman et al., 2012). Similarly, the massive wind power in Afghanistan and solar power potential in India can help the South Asian region go a long way in fulfilling its energy needs. And although India is aggressively pursuing its installed solar capacity, such initiatives in Pakistan and Bangladesh are far and few (except for the recently installed 100 MW solar center in Punjab, Pakistan). Intense research is needed to develop low-cost solar cells that could be used to generate power from sunlight for the entire South Asia.

As seen above, the climate change mitigation technologies (CCMT) is being pursued aggressively on global scale particularly in the field of energy generation. A total of 144 major patents have been filed worldwide on CCMT (Hascic et al., 2012). Of these, 12 patents have been filed to generate power from wind (9 on turbines, 3 on battery and grid connection), 23 patents on solar power (8 on solar thermal, 9 on solar cell, 4 on power conversion, and 2 on hybrid solar photovoltaic technology), 5 on geothermal energy, 4 on marine energy (1 each on OTEC, OWC, wave, oscillating water column), 5 on hydro energy (4 on dams-turbines, 1 on tidal/stream), 8 on biofuels (biodiesel, bioalcohol, ethanol), 4 from wastes (sludge, landfill gas, methane), 19 on efficient combustion (6 on heat and power, 13 on CO_2 mitigation from factories), 10 on capture, storage, sequestration, and disposal of GHG (bioseparation, chemical separation, adsorption, absorption, membrane, and condensation), 13 on energy storage (alkaline, lead-acid, lithium), 13 on hydrogen technology (production, storage, distribution), seventeen on fuel cells (membrane, methanol), 12 on emission abatement and fuel efficiency in transportation- electric vehicle, hybrid vehicle), 14 on energy efficiency in building (thermal insulation), and two on lighting bulbs (CFL and LED; Hascic et al., 2012).

Cooperation in developing CCMT is extremely important, as the research is cost-intensive. The patent data show that in all areas, a great deal of cooperation is involved between the US and Europe, followed by Belarus and Russia (in solar PV and thermal), India and the United States (solar PV, wind), South Africa and Europe (biofuels and wind), and the United States and China (solar PV). Maximum co-invention occurred in the field of CO_2 storage, followed by biofuels, CO_2 capture, solar PV, wind, hydro/marine, geothermal, and solar TH (Hascic et al., 2012). In this background, the South Asia could produce more than required energy from sunlight for which the entire region is especially blessed.

The International Solar Alliance (ISA) is an initiative of 121 sun-rich countries lying fully or partially between the Tropics of Cancer and Capricorn and receives 300+ days of sunshine throughout the year. The maiden summit of the ISA, cohosted by India and France was held in New Delhi on 11 March 2018. It pledged to generate 1 TW of solar energy by 2030, which would require US$1 trillion to achieve. The key focus areas of the alliance are promoting solar technologies, new business models and investment in the solar sector, formulate projects and programs to promote solar applications, develop innovative financial mechanisms to reduce cost of capital build, a common knowledge e-Portal to facilitate capacity building for promotion and absorption of solar technologies and R&D among member countries.

India contributed US$27 million to the ISA for building infrastructure and recurring expenditure over a 5-year duration from 2016–17 to 2020–21. France has committed for 700 million euros to the alliance. The ISA will have its secretariat in India. India will also provide 500 training slots for ISA member-countries and start a solar tech mission to lead R&D. India on her part has promised to generate 175 GW of clean and cheap electricity from renewable energy sources by 2022, including 100 GW from solar and 60 GW from wind.

5.3 CLIMATE INDUSTRY

The impact of climate change on the environment, society, and economy of the world probably lies in the strategy to develop climate industry that involves mitigation, adaptation, and engineering (Brasseur and Garnier, 2013). Mitigation involves drastic reduction in the emission of GHG at the global scale by decarbonizing energy generating system. Some of the mitigation measures were discussed in earlier sections, such as, the upcoming possibilities in terms of developing finer carbon neutral power generating technology, identifying probable areas of innovations and modifying human mindset. Adaptation, in contrast, involves developing comprehensive adjustment plans for different sectors. The climate engineering includes a spate of activities of relatively extreme nature that modifies the way solar radiation reaches the Earth or goes back to the space. Probably an efficient and harmonious integration of all three aforementioned strategies could save the Earth (Ahmed and Suphachalasai, 2014; FAO, 2012; Menzel, 2017; Tubiello, 2012).

5.3.1 Climate Change Mitigation

Climate change mitigation is a long-term measure aiming to reduce the amount of anthropogenic (human) emissions of GHG (Fisher et al., 2007;

Oppenheimer et al., 2014). Such reduction may be achieved by—(1) increasing the capacity of carbon-sinks through reforestation, and not allowing ocean to acidify, (2) improving the existing energy (electricity) generating units to low-carbon emission, (3) increasing insulation of buildings, (4) ensuring that new buildings use more of natural air and sunlight, (5) curbing the growth in demand of energy, and (6) stopping fiddling with the nature. The benefits of decisive and early action on mitigation may be more important and would cost less than that of adaptation (Fisher et al., 2007; Jacoby et al., 2014; Levine, 2007; Rahman et al., 2012).

South Asia experienced the highest average temperature rise since 2000, with Afghanistan recording maximum rise (1.57°C), followed by Bhutan and Pakistan (Kaur and Kaur, 2017). Agricultural lands, degraded soils, wetlands, and ocean are the best sinks to sequester (to hold within its structure) atmospheric carbon. The rate and intensity of such sequestration in soils depend on soil texture and structure, rainfall, temperature, farming system, and soil management. Interestingly in South Asia, the yield from these carbon-rich lands increases by 20–40 kg/ha (kilograms per hectare) for wheat, 10–20 kg/ha for maize, and 0.5–1 kg/ha for cowpea. Such sequestration could also reduce about 5%–15% of the global fossil fuel emissions to the atmosphere (Lal, 2004).

In this regard, mutual cooperation among South Asian countries could go a long way to mitigate the impact of humanity of this region. Such cooperation would avoid wasteful duplication of limited resources, and develop synergy for food security. To counter volatility in food prices, building regional food-stocks, food-godowns, and food-bank would be of much help (Kaur and Kaur, 2017). Also a 100-meter wide rail and road corridor linking Southeast Asia with Central Asia through South Asia may be set up expeditiously. This will not only improve communication, but would also improve trade and commerce, promote inclusive cultural growth and tolerance, and increase security and growth.

The Sustainable Development Mechanism (SDM) is an implementation tool of COP-21 agenda. The SDM is a successor-in-interest of Clean Development Mechanism of Kyoto protocol that got lapsed. SDM promotes GHG mitigation efforts over and above committed by respective countries under intended nationally determined contributions. Encouraging voluntary participation of nations, SDM insists for real, measurable, and long-term benefits related to the mitigation of climate change. The mechanism has provision to monitor, verify, and certify of each of these activities, using the experience gained with and lessons learned from existing market mechanisms (Prell, 2015). In other words, the SDM must—(1) contribute to overall emission reduction, that is, net mitigation, (2) account for mitigation targets of all countries including their progression over time, (3) encourage implementation of climate friendly policies, (4) promote low emission technology and policy landscape, and (5) support shifting from using fossil fuel to renewable energy.

Akhtar (2017) suggested that to achieve the sustainable development goals, South Asian countries must address demographic blessing of millions of youth by creating jobs, transform economy in a balanced manner, and undertake responsible industrialization. In addition, South Asian countries must accelerate infrastructure development, provide essential basic services to its entire population, including education, food, health and social protection, promote gender equality and women's entrepreneurship, and enhance environmental sustainability through low-carbon climate-resilient pathways

5.3.2 Climate Change Adaptation

Adaptation is an approach to save core assets or functions from the risks of climate change. Hence for its success, it needs ample care and empathy, and change in social mindset and political will (Pelling, 2011). This is because reduction in emission in global level in fact starts with change in behavior and mindset at household level. As the adaptive capacity of human being differs with geography and demography, the adaptation would require the situational assessment of sensitivity and vulnerability to environmental impacts (Green et al., 2009), which is linked to social and economic development (Galbraith, 2011; Schneider et al., 2007) afforestation.

Actually, compared to climate-change mitigation, climate-change adaptation policy development is still in its infancy. An ideal adaptive strategy (IFRC, 2009; Lwasa, 2015) must integrate adaptation into the long-term national sustainable development and poverty reduction strategies. Such an integration would strengthen existing capacities at the community, local governance, civil society, and the private sector levels, and develop robust resource mobilization mechanisms to ensure the flow of both financial and technical support to local actors. Setting up of early warning systems, contingency planning and integrated response to any calamity will be the other requirements of the adaptation strategy. The factors that constrain successful implementation of adaptation and mitigation options are given in Table 5.3.

With climate change affecting the pattern and intensity of rainfall, the South Asian soil could often show deficit in moisture. Such deficit would be harmful to the crop, particularly to that of wheat, soybeans, and corn. Hence, agricultural adaptation must include crop engineering (soil-quality-specific sowing of seed-type), arrange quick transportation to deliver surplus food to food-deficit areas, build smaller rainwater/river basin storage for irrigation, through revers interlinking popularize genetically modified crops after detailed research, and encourage watershed conservation and afforestation.

Table 5.3 Parameters Limiting Adaptation and Mitigation

Speed-breaking Parameters	Restraining Adaptation	Restraining Mitigation
Anomalous Population Growth and Rapid Urbanization	Pressure on natural resources and ecosystem services will increase	Growth in economy and demand in energy will emit more GHG
Less skilled intellectual manpower	Miscalculation on threat perception, funds requirement and adaptation approach	Reduce national, institutional and individual risk perception, willingness to change behavioral patterns and practices and to adopt social and technological innovations to reduce emissions
Divergences in social and cultural attitudes, values and behaviors	Reduce societal consensus regarding climate risk and therefore demand for specific adaptation policies and measures	Attitudes would impact on use of mitigation strategy and technology and thereby on pattern of emission
Challenges in governance and institutional arrange-ments	Reduce the ability to coordinate adaptation policies and measures and to deliver capacity to actors to plan and implement adaptation	Might negatively impact strategy and implementation mechanism for a carbon-neutral climate society, including using renewable energy
Lack of access to national and international climate finance	Reduces the scale of investment in adaptation policies and measures and therefore their effectiveness	Reduces the capacity of developed and, particularly, developing nations to pursue policies and technologies that reduce emissions.
Inadequate technology	Reduces the range of available adaptation options as well as their effectiveness in reducing or avoiding risk from increasing rates or magnitudes of climate change	Slows the rate at which society can reduce the carbon intensity of energy services and transition toward low-carbon and carbon-neutral technologies
Insufficient quality and/or quantity of natural resources	Reduce the coping range of actors, vulnerability to non-climatic factors and potential competition for resources that enhances vulnerability	Reduce the long-term sustainability of different energy technologies
Adaptation and development deficits	Increase vulnerability to current climate variability as well as future climate change	Reduce mitigative capacity and undermine international cooperative efforts on climate owing to a contentious legacy of cooperation on development
Inequality	Would put disproportionate pressure on male and female and on most vulnerable sections to adapt differently	Constrains the ability for developing nations with low income levels, or different communities or sectors within nations, to contribute to greenhouse gas mitigation

Source: Modified *from Pachauri et al. (2014): Synthesis Report. Contribution of Working Groups I, II and III to the Fifth Assessment Report of the Intergovernmental Panel on Climate Change (Core Writing Team, Pachauri, R.K. and Meyer, L. (Eds.)). IPCC, Geneva, Switzerland.*

Adaptation can also collaterally help in reducing poverty, improve in health care, spreading of education and awareness, and in the process that could promote sustainable development (Adger et al., 2007; IPCC, 2014). However, one should think of long-term adaptation measures to avoid being short-circuited. Similarly, adaptations at one scale may reduce the adaptive capacity of others. National adaptation programs of various South Asian countries are interesting- while India constructs embankments to stop the flow of surface water and run off, Pakistan and Nepal are improving livestock rearing productivity, and Bhutan is going for drip irrigation using bamboo. Decline of public investments in agriculture sector in South Asia is disappointing. For example, in Bhutan investment declined from over 30% in 1980 to under 10% in 2015. Water security is also a contentious issue. Dwindling of water for irrigation, particularly in arid and semiarid regions is estimated to increase by at least 10% if temperature goes up by 1 °C (Laborde, 2011).

In summary, climate change adaptation must be able to sustain principally three things—human health, agricultural yield, and economic development. Role of adaptation becomes more important because even if GHG emissions are stabilized relatively soon, global warming and its effects will last for many years, and adaptation will be necessary even to the resulting changes in climate (Farber, 2007).

5.3.3 Climate Engineering

Climate engineering envisions developing appropriate and carbon-neutral technologies to reduce the intensity of climate change. The technology may either capture a considerable amount of CO_2 present in the atmosphere or generate a physical mechanism at the planetary scale that would compensate for greenhouse warming (NAS, 2015 a, b).

While being involved in climate change issues, one could find super-environmentalists on one side and climate engineers on the opposite side. For example, the role played by some of the environmentalists is conspicuous. According to Idso and Singer (2010), the environmentalists seem to be more worried with generation of wind energy because the turbines might kill bats. They are equally anxious with solar energy that lizards may die in the desert, and much bothered with geothermal power plants because earthquake might damage the geothermal unit if it is built over a fault plane. These environmentalists are also apprehensive that fish might get caught in the turbines of tidal energy and similarly with hydro energy because it may pollute river water. However, these environmentalists have no concern for pollution created by oil spills from oil drilling, natural gas extraction, coal extraction, and the poisoning and polluting of water and land caused by the nuclear radiation storage (Idso and Singer, 2010; Idso et al., 2014).

Climate engineers, in contrast, represent the other extremes. It goes much beyond the mitigation and adaptation (see earlier sections). Climate engineering aims to design new technologies to effect large-scale manipulation of planetary environment to alter global climate by reducing undesired anthropogenic climate change (Keith, 2000; Shepherd, 2009). Climate engineering can be achieved through two processes—solar radiation management (SRM, altering the Albedo, reflectivity of Earth) and carbon dioxide removal (CDR, i.e., removing carbon from the atmosphere). SRM aims to counteract global warming by reflecting solar radiation to space, whereas CDR aims to remove CO_2 from the atmosphere (Table 5.4).

SRM (reducing short-wave incoming radiation to the Earth) may use a host of cheap, sensible surface-based activities. For instance, altering the ocean surface brightness and modifying the roofing material of building could reflect back solar radiation efficiently. Similar way, spraying seawater over clouds brightens the cloud-surface and act as effective reflector. Similarly growing high-albedo crops could return the solar radiation back to the Space. Induction of highly reflective sulphate aerosols in the upper atmosphere and creation of space-sunshade using mirrors and dust are some of the climate engineering measures (Bewick et al., 2012; Keith, 2010).

The CDR methods, in contrast, involve (Ciais et al., 2013; Ming et al., 2014)—(1) afforestation/reforestation, (2) improved forest management, (3) sequestration of wood in buildings, (4) biomass burial, (5) biochar, (6) conservation agriculture, (7) fertilization of land plants, (8) creation of wet lands, (9) biomass energy with carbon capture storage, (10) algae forming and burial, (11) blue carbon (mangrove, kelp farming), (12) modifying ocean upwelling to bring nutrients from deep ocean to surface ocean (ocean fertilization), (13) enhance weathering overland, and (14) enhanced weathering over ocean. Ocean fertilization involved dumping of raw iron pieces on seabed to increase phytoplankton growth and production that would in turn sequester CO_2

Table 5.4 Mechanisms for Climate Engineering

CDR Technologies	SRM Technologies
• Bioenergy with carbon capture and storage • Afforestation • Biological carbon sequestration • Direct air capture • Accelerated chemical weathering • Ocean iron fertilization	• Particle injection • Cloud brightening • Increase deflection • Planetary shading

CDR, *Carbon dioxide removal;* SRM, *solar radiation management.*
Source: *Modified after Gordon (2017).*

(Boyd et al., 2007). This experiment however, caused substantial environmental controversy. In addition, accelerated chemical weathering of rocks, manufacture of products using silicate rocks and carbon from the air, and direct capture of CO_2 from the air are other mechanisms for CDR (Caldeira et al., 2013; IPCC, 2013).

Climate engineering, as seen by many, is all about artificial intelligence to fiddle with nature in a reckless pursuit to neutralize climate change, without realizing probably that these actions might be boomeranged on some other areas of the globe, at some other time and at disproportionate intensity. Moreover, the scientists are divided with regard to the moral and ethical level of climate engineering that involves nanotechnology and genetic engineering (CEC, 2017). Because climate engineering represents a large-scale, international effort to modify the climate artificially, it raises questions of whether humans have the right to change the climate deliberately, and under what conditions (Appell, 2008).

Meanwhile, meaningful research and development of newer products to reduce carbon emission are no more a choice; rather it is an imperative, as demand for energy was increasing even a decade back at the rate of only 2%–3%, now it is at 20%–30%. Replacing petroleum and diesel with natural gas in the transportation sector has been a great boost. In decades to come, scientists will begin to master combined uses of molecular biology, nano-engineering, and robotics in climate engineering. In fact, engineers will face challenges to replace twentieth century infrastructure—roads, power lines, pipelines, and so forth, with new technologies and materials.

Along this trend, climate engineering may also include small reforms (change in our behavior) that could do wonders. For example, *Mobiliteam* was developed to boost air that reduces the energy consumption of electric vehicles by improving the efficiency of air conditioning systems. Similarly, *Bynd* is an app facility that facilitates nearby workers to combine journeys and in the process saves fuel consumption and GHG emission. *Traffic energy bar system* is used where high volume of slow moving traffic press down bars on the road by the wheels of each car as it moves over them, creating an up and down motion that generates electricity. *Mutum* aims to encourage sharing, discourage overconsumption, and avoid wasteful industrial excess manufacturing (CEC, 2017; Gordon, 2017).

In summary, it may be recommended that the world should make increased efforts toward mitigating and adapting to climate change, while advanced research on low-cost, low-risk, and low-disruptive climate engineering mechanisms must continue (Royal Society, 2009). While assessing the CDR methods that accelerate the removal of atmospheric CO_2, and the SRM measures that reduce solar absorptions by about 2%, Bala (2011) insisted that prevention (mitigation) is better than cure (climate engineering).

CHAPTER 6

Alternate Governance Policy

6.1 EVOLUTION OF GOVERNANCE

South Asia has a rich history of governance. The first thought of inclusive governance was prophesied during Lord Rama's empire (Ramayana era, ~5100 BC). Later a polished and much constrained philosophy of governance was made available during Mahabharata era (~3067 BC). In both these cases powerful Planetarium software was used by BN Achar and his team to arrive at the dates (Achar, 2005). They recorded the days (*tithis*) according to the star (*nakshatra*) on which the moon prevailed. Added to these were the months, the seasons, and even the different Solstices. They calculated the days of events by noting a particular arrangement of the astronomical bodies, which occur probably once in many thousand years.

However, probably the first structured, spiritually inspired, and scientifically perceived governance mechanism was proposed by the great minister Chanakya (also known as Kautilya) during the reign of King Chandragupta Maurya (317–293 BC, i.e., 2335 to 2311 years before present). This was done in spite of his constraint in technology, communication, and infrastructure facility (Jamil et al., 2013). Kautilya's *ArtháShastra* was a science of politics intended to teach a king how to govern. Minister Chanakya offers wide-ranging insights on war and diplomacy, on natural allies and inevitable enemies, on geopolitics and using secret agents, and on the use of religion and superstition to bolster his troops and demoralize enemy soldiers. At the same time, he had given instruction for humane treatment to the conquered soldiers and subjects Boesche (2003).

The basic steps considered to achieve good governance have been critical accountability on the part of the government, transparency in decision making, and being participatory and inclusive in the implementation of decisions (Fig. 6.1). The governance should remain responsive (sensitive) to the problems around and must try to develop consensus in arriving at a decision. Bhattachrjee (2011) claimed that the modern management paradigms were

Climate Change. http://dx.doi.org/10.1016/B978-0-12-812164-1.00006-2

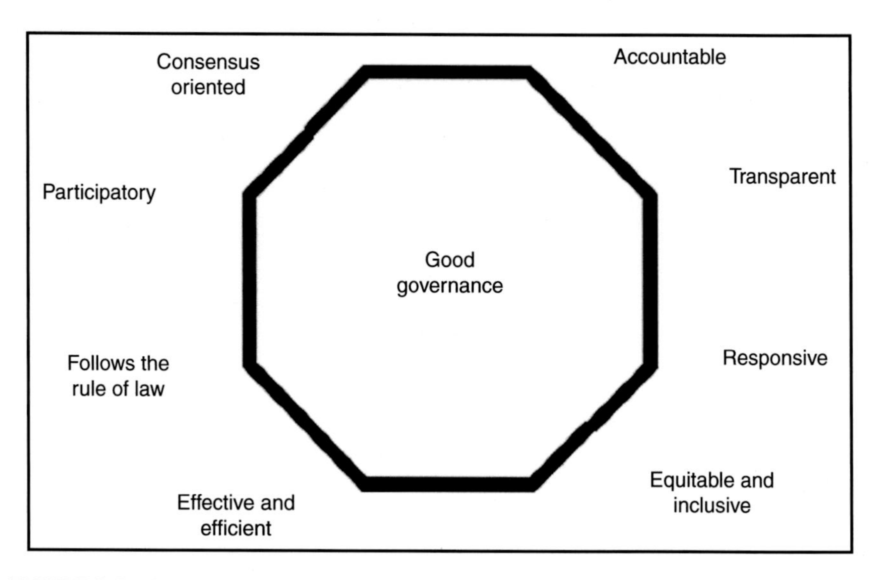

FIGURE 6.1 Characteristics of good governance. *(From United Nations Economic and Social Commission for Asia and the Pacific, What is Good Governance? ©United Nations. Reprinted with the permission of the United Nations)*

derived essentially from the ethical and spiritual frameworks of ancient Indian ethos and scriptures like Vedas, Upanishads, Bhagavad-Gita, Manu-Smriti, and Arthashastra.

The noted philosopher the late Mohandas Karamchand Gandhi, while speaking on austerity once articulated that this "Earth has everything in abundance for the need of all persons, but not for their greed." Today, Gandhi's insight is being put to the test as never before. Satpathy et al. (2013) found that the UNESCAP's (United Nations Economic and Social Commission for Asia and Pacific) norms of good governance are closely related to Bhagavad-Gita's (the ancient Indian religious text) intrinsic perspective of governance. Such norms have the potential to take self-governance to the level of corporate governance and elevating later to global governance status.

The need for inclusion of spiritual understanding in governance was argued earlier (Cheung, 2005; Jamil et al., 2013). In fact, imbalances in social, political, economic, and military dispensation have resulted in global warming, climate change, and loss in biodiversity. The impact of such disparity, however, is felt differently by rich and poor nations. Hence, any effective governance must also include the way power and wealth are globally distributed. Bringing in change attitudes and behavior of all rich and poor alike is therefore essential to protect and conserve the environment and to promote peace. The existing way to live in a selfish world, separated from nature, and using the Earth's limited resources to satisfy our relentless unending greed, must end rapidly.

Table 6.1 Economic growth of South Asian countries

Countries	Real GDP growth (%)		
	2016	2017	2018
Afghanistan	2.0	3.0	4.3
Bangladesh	7.1	6.8	6.5
Bhutan	6.4	6.6	7.0
India	7.1	7.1	7.5
Maldives	3.9	4.5	4.6
Nepal	4.6	4.6	4.8
Pakistan	4.7	5.2	5.4
Sri Lanka	4.4	4.8	4.9

Source: Modified from *United Nations ESCAP (Economic and Social Commission for Asia and the Pacific) Report on national sources (2017)*

Growing at 4%–5% annually, the world economy will be on a path to double its size in less than 20 years from now. Today's $70 trillion world economy will be at $140 trillion before 2030, and $280 trillion before 2050, if we extrapolate considering today's growth rate (UN-ESCAP, 2017). The gross domestic products (GDPs) of South Asian countries (SAC) are also strengthening (Table 6.1), which over the last two years has been more than doubled in Afghanistan, and so also in most other SAC. Similarly the average per capita income is $10,000, with the rich world averaging around $40,000 and the developing world around $4000. However, as if to neutralize the economic growth, population in the globe is increasing fast. Compared to just three billion half a century ago on this planet, the footfall has now reached to seven billion people. In both the cases, our planet will be unable to physically support this exponential economic growth if we allow greed take the upper hand. So also, the world will be unable to feed 9–10 billion people (detail discussed in page 148). A calamity is inevitable unless we change.

In this background, the human kind has now three paths to follow: (a) Capital Economy that emphasizes profit and requires perpetual growth, (b) Religious Economy that divides the world into narrow religious bigot economy, or (c) Ethical Economy that provides a value-based spiritually inspired balanced standard of living in complete harmony with the biosphere, hydrosphere, and atmosphere. Ethical economy argues to put the long-term goodness of humankind above the short-term benefit, and to put carbon level in atmosphere back to 350 ppm from today's around 400+ ppm.

6.2 THE APE CONUNDRUM

As money mints money, power wins greater power and wealth for those residing at the top of the society. Consequently, as mentioned earlier, the world has witnessed an apparent positive relation between greenhouse gas (GHG) emission and strategic power index (SPI) of a nation, in a way similar to the gross domestic product (GDP) of a country reflecting its economic strength. It is because prosperity needs energy, and GDP closely correlates to such prosperity (Mukhopadhyay, 2018). One percent economic growth means one percent more emission of carbon. Thus, a decrease in CO_2 emission may result in a commensurate decline in prosperity.

It is further seen that the country, which has enhanced energy production appear to record higher GDP. Consequently these countries would achieve two-to-three fold prosperity, would become more powerful economically, and could strategically influence regional and global affairs. Now which strategically powerful country on this Earth would like to lose this power and status? Hence, a fundamental conflict appears to exist between increasing prosperity (read aspiration and performance) and reduction in CO_2.

The changes in global temperature and consequently on climate will impact a region like South Asia more, which holds about 1.7 billion people (~23% of world population) in an area of about 5.1 million km^2 (~ 11.51% of the Asian and 3.4% of the world's land surface area; Sikdar and Mukhopadhyay, 2016). In fact, approximately 30%–40% of the world's poor live in rural regions of South Asia. While the developed countries may talk about environment, the developing South Asian Countries (SAC) countries, are busy evolving a strategy to get their 1 billion poor people in to the middle class. It does not matter whether they have carbon emitting coal power plant or whether the labor standards in the coal mines were low, - the way it did not matter for the developed countries in the 19th century. Naturally, SAC, particularly India, would not like to decouple development from using coal (and CO_2 emission from coal), if so needed.

In this regard, the ever rising deficit between aspirations of people to live a high-end life, vis-a-vis optimum utilization of natural and manufactured resources (that reflects integrated performance of the system) makes things worse. The Aspiration-Performance-Ethics (APE) relationship in the background of increasing population and augmented carbon emission was studied by Mukhopadhyay (2018).

Aspirations are vital to the growth of a human being and also for the society. It keeps the clock of development moving. However to achieve rational aspirations one has to have adequate resources—natural, manufactured, educational, skill, finance, and managerial. All-round performance to make best integrated use of available resources will be required. However, there is every possibility of such campaign going (and probably humanizing way) haywire if aspiration and performance are left as they are. The way any chemical (fission) reaction needs regulator, the role of ethics in normalizing speed, direction, and quality of fulfillment of human desire was emphasized by Mukhopadhyay (2018). The answer probably lies in focusing on increased investment in human capital (education and health), and steps to make development efforts environmentally responsible (Langmuir and Broecker, 2012; Mesagan, 2015).

With the income and profit influencing the decisions and behavior in almost all industries, ethics normally takes a backseat. These industries view the world in terms of a game like soccer, wherein you set out to defeat opponents unmindful of situations that you may also receive goals. This 'soccer-attitude' at times transforms legitimate profit into greed, tempts to cut-corners, and bulldozes the fair level playing field. In spite of this attitude and formidable industrial competition, the importance of ethics and its relationship with business success are still being well-guarded.

Immanuel Kant, the 18th century ethical philosopher defined that doing business ethically is challenging as conflict of interest occurs between what seems tangible benefit and what seems responsibility. But it seems that ethics and success in business need not necessarily be in contrary, but certainly complex (Martin, 2010). The art to balance the duo (aspiration and performance) with a clear purpose of growth has always been challenging.

Mukhopadhyay (2018) used eight data sets to examine the interrelationship among peoples' aspiration, performance, and ethical approach. The indices of happiness (HI), human development (HDI), gross domestic product (GDP-PPP), ease of doing business (EDB), gender equality (GEI), and corruption perception (CPI) in the background of increase in carbon emission and anomalous population growth were examined (Table 6.2). All these information from 22 countries from different economic and development stature (8 South Asian, 7 OECD [Organisation of Economic Cooperation and Development], 4 other leading Asian, and rest 3 from fast developing countries) were included for the study.

APE in relation to concentration of carbon in atmosphere has been interesting. One of the most performing asset to act as the "sink" for atmospheric carbon dioxide has been the ocean, which absorbs about 30%–40% of atmospheric CO_2 (Hardy, 2003). Increased CO_2 absorption is turning the ocean more acidic, thereby imposing a threat to the entire marine life. Another important factor has been the increase in population. Using even moderate to conservative demographic projections, the world's population will almost reach 9 billion by mid-21st century from 7 billion at present, against a sustainable carrying capacity of 3–4 billion by this Earth (HDR, 2016; WRI, 2015).

Anomalous growth of population is a product of poverty, illiteracy, diseases, malnutrition, gender discrimination and lack of women empowerment. In a circular reversal, the growth in population creates all such social viruses. Poverty can be alleviated through development of the entire population, instead of supporting a section of the society. Without concrete measures for growth

Table 6.2 APE data from OECD, South Asia, and few other major countries

Country	Aspiration		Performance		Ethics	
	HI	HDI	GDP-PPP	EDB	GEI	CPI
Norway	7.537	0.949	70,666	8	0.053	85
Canada	7.316	0.920	47,771	18	0.098	82
Australia	7.284	0.743	50,817	14	0.120	79
USA	6.993	0.920	59,609	6	0.203	74
Germany	6.951	0.926	49,815	20	0.066	81
UK	6.714	0.910	44,001	7	0.131	81
UAE	6.648	0.840	68,425	21	0.232	66
Brazil	6.635	0.754	15,485	125	0.414	40
France	6.442	0.897	43,653	31	0.102	69
Russia	5.963	0.804	27,466	41	0.271	29
Japan	5.920	0.903	42,860	34	0.116	70
China	5.273	0.736	16,676	78	0.164	40
Pakistan	5.269	0.550	5,375	147	0.546	32
Nigeria	5.074	0.353	5,960	145	-	35
Bhutan	5.011	0.607	4,578	75	0.477	65
Nepal	4.962	0.558	2,642	105	0.497	29
S Africa	4.829	0.666	13,409	82	0.394	45
Bangladesh	4.608	0.579	4,207	177	0.520	26
Sri Lanka	6.440	0.766	13,012	111	0.386	36
India	4.314	0.624	7,153	100	0.530	40
Afghanistan	3.794	0.479	1,833	183	0.667	15
Maldives	-	0.701	16,276	136	0.312	36

Source: Modified from *COMEST (2010) Happiness Index, Human Development Index, Corruption Perception Index (Transparency International Corruption perceptions index, pp.1–9, 2016. www.transparency.org), GDP Per capita-Purchasing Power Potential 2016 in US$; Gender equality Index (hdr.undp.org, 2016), Ease of Doing Business (2017), More the better for HI, HDI, GDP-PPP, and CPI, whereas less the better for EDB and GEI, https://data.worldbank.org/indicator//.*

and poverty eradication, other methods of population control may prove to be ineffective. Hence, government policies should be deeply entrenched to influence the economy of the state and the daily life of the people, and hence needs to be science based. However, such rational doctrine is not followed in many situations in South Asia owing to political and social considerations.

The most significant interrelationships have been found between HDI and GEI, and EDB and CPI ($R^2 > 0.75$; Fig. 6.2). These relationships suggest that gender equality is must-a-tool toward human development, so also a transparent business environment for augmenting trade and commerce. The second rung of significant relation ($R^2 = 7.0–7.5$) exists between gross domestic product and four other parameters (HI, EDB, GEI, and CPI). Although the exact causal relations between GDP and two other parameters—EDB and GEI—is yet to be ascertained with certainty, their strong positive relations might indicate that if a country is interested to increase its GDP, she cannot probably remain oblivion to ease of doing business and gender equality, or vice-versa

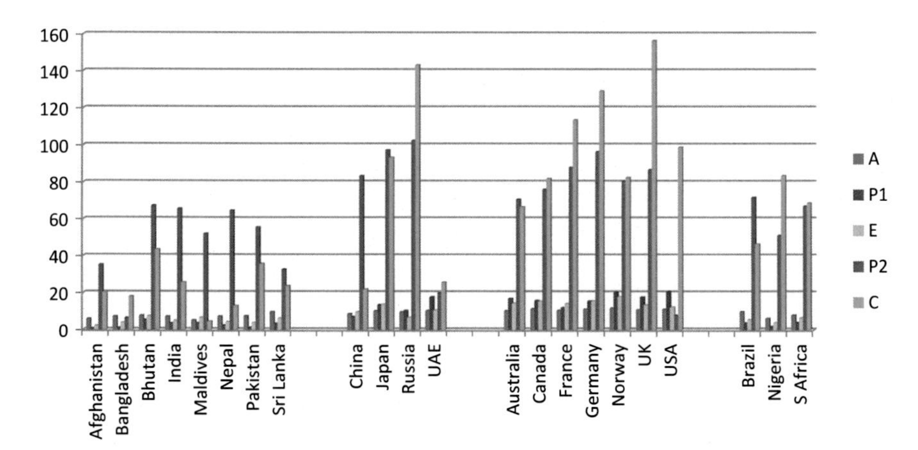

FIGURE 6.2 Average aspiration (A), performance (P1), ethics (E), percent increase in population (P2) and carbon emission (C) for all countries.

(i.e., if a nation has high GEI and EDB, it can expect a rise in GDP). Similarly, as GDP directly affects the economic and consequent mental condition of the population, the GDP and HI show very positive relation (R^2 = 0.719). A transparent and corruption less business environment (CPI) would also boost GDP (Mukhopadhyay, 2018).

A collage of all essential parameters (aspiration, performance, ethics, in the light of carbon emission, and population) for these 22 countries evinced interesting impressions (Fig. 6.2; Table 6.2). Among the SAC, the ethical values are maximum in Bhutan followed by Maldives and Sri Lanka. Whereas, Afghanistan is having least ethical values which is also understandable from its socioeconomic conditions. Bhutan is also maintaining a good balance of aspiration, performance, and ethical values. Probably due to moderate to high ethical values, Sri Lanka also would be able to fulfill a fair degree of aspirations in spite of having scarcity of resources and low level of performance. Similar is the case for Maldives, Brazil, and South Africa. The noticeable thing is that in spite of having very high performance and ethical values, the achieved aspirations of the developed Asian Countries (China, Russia, Japan, and UAE) and OECD countries are not able to go beyond a certain limit.

In this regard, one fact must be kept in mind that in SAC the wealth is not well distributed and only few people have a chunk share of the entire economy giving rise to the high (distorted?) GDP. For example, even with low in performance and ethical values India is having third rank in GDP, which is driven by only few people and incorrectly represents the entire population.

Depicting the social dynamics in South Asia, the study suggests that an ethical aspiration could take best care of the performance crunch while maintaining responsible level of aspiration in this part of the world. The present mindset to consume like mad, borrow like crazy, and believing that a high GDP can buy peace, happiness, and harmony needs to be revisited. While humanity must continue to progress, the effort should also focus on developing a socially-just-society where all individuals are able to flourish. Over-simplification of the importance and significance of culture and ethics in dealing with infrastructure development could be catastrophic.

6.3 BTG DOCTRINE

Realizing that the time has come to repair the deficit of trust and goodness in the society by removing economic and social disparity, and by ensuring inclusive growth, leading to the creation of balanced human being, it is now increasingly realized that the progress made in the fields of science and material-products need to be complemented with equal intensity of ethics, values, and egalitarianism. Such an initiative would enable sustainable, balanced, and purposeful growth in the society. For example, in mitigating the climate change impact, if innovations and governance are considered as the hardware, then mindset is certainly the software. To make public administration successful, visible, and relevant, the mindset of the individual and society needs moderation. And here is where a new doctrine of governance must come in. South Asia in this 21st century and beyond probably needs a governance strategy where man and nature will remain close entities.

In this direction, the BTG doctrine was introduced (Mukhopadhyay, 2018). This doctrine is built over the philosophy and teachings of spiritual guru Lord **B**uddha, Nobel laureate writer Rabindranath **T**agore, and social philosopher Mohandas Karamchand **G**andhi (BTG). The implicit doctrine of harmony between human and nature as prophesied by these three best known philosophers could play a vital role in mitigating, adapting, and engineering the climate change impact. The BTG doctrine explains that what the humanity is facing today vis-à-vis climate change is nothing but ecological consequences of their collective act (*karma*).

According to this doctrine, if greed dominates, the engine of economic growth will fast deplete the resources, push the poor aside, and drive people into social, political, and economic doldrums. It suggests political and social cooperation, both within the countries and internationally. There will be enough resources and prosperity, if one learns to share. The greed to consume more and more is an expression of craving, the very thing the BTG pinpoints as the root cause of suffering. In fact, our present economic and technological relationships with the rest of the biosphere are unsustainable. Hence, the present generation and policymakers must listen to the silence of nature and of those who cannot express, and must be their voice too, and act on their behalf. Social awakening in this regard has to arrive soon.

Trees and forests (read nature) have been very special to Buddha, Tagore, and Gandhi. It was under a tree that Buddha was born in Lumbini (Nepal), achieved enlightenment, and passed away in Bihar (India). Tagore embarked upon open-sky under-tree education in his University in Santiniketan (Bengal, India). Gandhi spent his entire life in Sabarmati Ashram (Gujarat, India) amidst nature and miles away from Capitalist economy. If carbon credit deal is meant to essentially offer a reward for protecting the forest, then by one estimate, the credits for all 13 forests planted and nurtured by monks in Cambodia could be worth as much as 50 million dollars over 30 years (COMEST, 2010).

The BTG doctrine suggests that humanity can only excel when it gets strongly connected to nature (Fig. 6.3; Table 6.3). Such intimacy may grow from

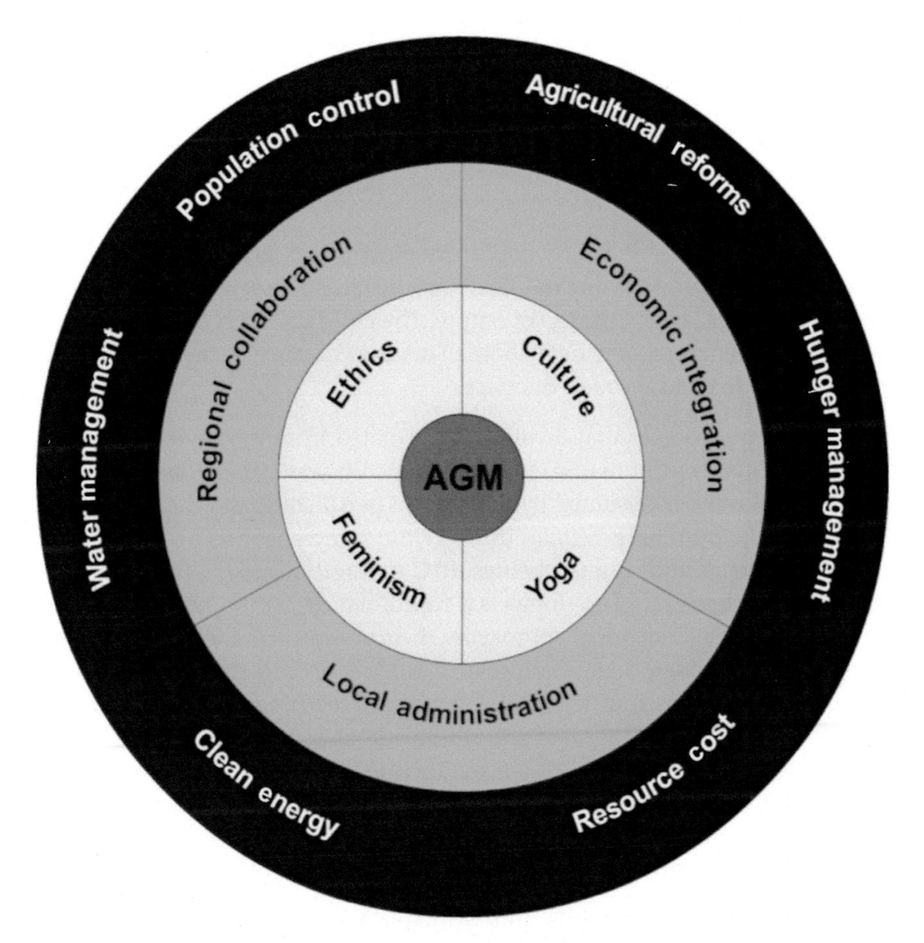

FIGURE 6.3 The BTG wheel of alternate governance model (AGM).
Outer *blue* layer—implementing action areas, intermediate *green* layer—mechanisms to create trust and security, inner y*ellow* layer—core aspects of to change mindset; *AGM*, Alternate Governance Model.

Table 6.3 BTG (Buddha-Tagore-Gandhi) Doctrine as a mode of alternate governance

Layers	Components	Remarks
Soul + Mindset (inner, yellow)	Ethics (honesty, transparency), culture (poverty, plurality), feminism, and yoga (human development)	Buddha-Tagore-Gandhi Doctrine insists for an optimum mixture of science, technology, spiritualism, and nature to alleviate GWCC threat through ethical adaptive governance.
Strategy + Attitude (Intermediate, Green)	Regional collaboration and economic integration Local administration and participatory governance	Cooperation among South Asian nations is must. Sharing resources and hassle-free trade are perquisite. Empower local government.
Mitigation + Adaptation (Outer, Blue)	Population control Water management Agricultural reforms Clean energy Resource cost Hunger management	These are the actions to be taken up immediately in a time-bound manner with all earnest and dedication, without any favor, coercion, or indifference

childhood, as parents and early educators must play a major role. Instead of teaching within four walls, the toddlers must be taught under open sky beneath the shade of trees and on the lap of nature. The child will then be fascinated by the splendor of nature—the rising of the Sun, the chirping of the birds, and the whistling of the wind through the trees.

The BTG doctrine argues that divine is not isolated from the world. Rather the infinite personality of human comprehends the universe. There cannot be anything that cannot be subsumed by the human personality, and this proves that the truth of the universe is human truth. While some scientists believe that the truth is independent of human beings, BTG doctrine suggests that truth is the perfect comprehension of the universal mind, which can be realized only by human beings. The individual approaches the truth through their own mistakes and blunders, through their accumulated experience, and sometimes through their illuminated consciousness. The summarized teachings of Buddha, Tagore, and Gandhi suggest integration of philosophy into the study of good governance. The BTG Doctrine the (Wheel of Governance, Fig. 6.3) has three layer-components that constitute Alternate Governance Model (AGM). The components include soul-mindset along inner yellow layer; strategy-attitude in intermediate green layer; and action on mitigation-adaptation in outer blue layer (Table 6.3).

BTG combines a set of simple and workable remedial measures, which in turn coalesce science, technology, nature, spiritualism, and commonsense to bring about a change in our thoughts, understanding, values, and the way we perceive our existence. The actual dispensation of the BTG wheel of ethical governance (Doctrine) on the ground is shown in Table 6.4.

6.3.1 AGM: Inner Yellow Layer: Soul and Mindset

It is now increasingly felt that fight against impact of climate change cannot be completed and successful without addressing the soul and mindset, which comprises of ethics, culture, feminism, values, education, literacy, freedom, justice, transparency, nonviolence, equality, and finer points of Yoga. Northcott (2007) based on moral implications of climate science suggested that nature and culture, and science and ethics are inextricably linked. Also it underwrites that those who suffer most from the weather extremes in the less industrialized nations have the least responsibility for their situation while those who have most benefited from the wealth generated by fossil fuels suffer the least. We briefly discuss below the role of ethics, culture, feminism, and yoga in developing a harmonious society.

Climate change needs a global solution in an apparently imperfect world. And besides science, the other aspects like fairness, equality, and justice could play major roles in mitigating such change. Somerville (2008) identified three major ethical dilemma—how to balance the rights and responsibilities of the developed and developing world; how to evaluate geo-engineering schemes designed to reverse or slow the rate of climate change; and how to assess our responsibility to future generations who must live with a climate we are shaping today. These questions emerge as to ask who has given geo-engineers the moral and legal right to tinker with the global environment. Again, arguing that climate policy should be driven solely by national self-interest is hugely unethical, as it is a global problem. Immoral mitigation approaches might set things right locally, but would miss longer-term broader concerns (Gardiner, 2016).

Culture: Culture is often described as a comprehensive knowledge that human beings require to function in a certain situations. These are knowledge of language, habits, rituals, opinions, values, and norms (Hofstede, 1994; Shadid, 2007). Strauss (2012) concluded that cultures are not static and do not exist in isolation, rather evolve with time in response to changes in surroundings (wars, plagues, new inventions), as well as to the environmental and climate variability.

Adger et al. (2013) argued that society's response to every dimension of global climate change is mediated by culture. Bhattacharya and Wertz (2012) while narrating the issue of climate change, migration, and conflict, opine that with high population density in South Asia, the climate change has the potential to create complex environmental, humanitarian, and security challenges.

Visweswaran (2004) believes that culture to a great extent, builds over human rights conditions and the culture in South Asia, shows gender-biasness and is violently masculinized. The dowry deaths in India, honor killing in Pakistan, acid-throwing incidents in Bangladesh, and ethnic-religious skirmishes in Afghanistan and Pakistan are some of the "He-attitude" examples. The governments of SAC must move out of the existing mindset which considers an adversarial relationship with a neighboring country as the emblem of patriotism; affluence of the few at the expense of the many as the hallmark of development; individual greed as the basis of public action; and mistrust as the basis of interstate relations (Hussain, 2006).

Feminism: The climate change impacts on the women, who are considered to be disproportionately, affected owing probably to patriarchal or hierarchical structures prevailing in the South Asian societies. The saying goes—Men pollute, and Women suffer. The "He-image" is usually associated with power, strength, domination, aggression, meat-eating, and car driving. All these emit carbon to the atmosphere. Breaking this patriarchal, conventional, and traditional concept not only could bring parity but would also reduce emission.

As the livelihoods of women are more dependent on natural resources, women are further threatened by climate change (World Bank, 2015). The Paris agreement realized that the fight against the impact of climate change has given this world a unique opportunity to focus on gender equality. The COP-21 signatory-countries are to identify gender-sensitive strategies to respond to the environmental and humanitarian crises caused by climate change. Integrating gender into climate policy will be an elegant policymaking. Such an approach might eventually help neutralize the discrimination in terms of gender, education, income, and age.

With restricted mobility, disproportionate family set-up, unequal property arrangement, and discriminatory access to both resources and decision-making processes, women in South Asia are most vulnerable to climate change impact. For example, women have to walk farther to fetch water and firewood owing to climate change, which might develop some health problems in them in future (Sousounis, 2016). Unfortunately, women are unable to voice their specific requirements either at the local project level or in international negotiations (Bäthge, 2010).

As capitalism was alleged to have founded on the subordination of women and nature, for some, climate change mitigation is a feminist, antiracial and depatriarchal issue. In fact, every activity of women (choice of voting, choice of purchasing goods, choice of sex, choice to conceive) can constitute a moral–political dispensation. The bottom line is that for a healthy Earth, women must have healthy living conditions starting from the land and the environments they rely on for food, water, health, and everyday life (Awadalla et al., 2015).

Proposed withdrawal of USA from the Paris Agreement could also be seen from patriarchal angle. The entire debate got further vitiated with climate change is conveniently clubbed with nationalism, knowing well that GWCC cannot be solved at a national level. A sustainable long-term management of water resources would probably require more of women's participation, because women have specialized knowledge crucial for coping with drought, floods, and damage due to extreme weather (Perkins, 2014). The feminist political ecology to frame watershed-based women centered community engagement could provide sustainability and respectability to the program in South Asia.

Moreover, the extensive theoretical and practical knowledge of women about the environment and resource conservation is not given due consideration, and their potential contribution to climate mitigation by being part of the economic cycle is not sufficiently exploited (Bäthge, 2010). To make women counted in GWCC mitigation and adaptation issues, promoting women's economic participation and empowerment is a must. Such an involvement would also foster economic growth, social development, and poverty alleviation. Hence to win the battle against the threat of GWCC in South Asia, one need not only to be innovative, but will be also asked to bring about a change in mindset, that encourages gender equality, and gender empowerment.

The perceived divide between religion and science has the potential to essentially alter the climate debate (Hayhoe, 2015). Pope John Paul in June 2015 spoke eloquently calling 2 billion Christians in the world to take lead to keep God's creation of this humanity safe. Similar stewardship is suggested also by the Islamic declaration in August 2015, which connects the issue of climate change to humans' relentless pursuit of economic growth and consumption. It encouraged the world's 1.6 billion Muslims to support global action on climate (Hayhoe, 2015).

Yoga-way-of-life: Another large community—the Hindus (more than 1 billion population) has always been prescribing Yoga-way-of-life of containment, wherein humanity should live for its need, and not for greed. The biggest problem the world is facing today, after terrorism, has been the climate change. In fact, solution to both these problems (terrorism and climate change) could probably be found in Yoga. Yoga integrates various forcing parameters that try to influence human being, and is a complete training cum spiritual exercise of body, mind, and soul. Yoga, it is said, offers the complete comprehensive knowledge, without compartmentalizing various aspects, almost the similar way several issues in struggle against GWCC are interlinked (Boas, 1928; Strauss, 2012).

The first lesson of yoga is—"be the change yourself to recognize and embody your unity with the environment, and through this recognition, to change your personal practices." Relate individual smallness to vastness of nature and natural environment, as human beings are not separated from nature. Yoga offers people a sense of "harmony with self, society, and nature" and could create a "social consciousness." Time has come to discuss beyond rate and time of emission cut, and search for sustainable solution.

Yoga brings about the much needed peace within oneself. It balances one's existence, relates her/him to the nature, and offers him/her the ability to appreciate all the dynamic forces in this world. The urge to gather unnecessary resources at the expense of other's legitimate share will vanish. Such a situation will help people to get satisfied with need and not greed. The teachings of Yoga could make one strong from inside so much that any undue aspiration for material benefit will not emerge in his/her mind.

6.3.2 AGM: Intermediate Green Layer: Strategy and Attitude

An ethically and culturally pure soul and a gender-neutral mindset (Section 6.3.1) must guide attitude, thinking faculties, and strategies of a human being or nation to implement mitigation and adaptation mechanisms against the impact of climate change. Climate change is a challenge, as well as an opportunity. If so, enhancing the resilience and capacity of the society to respond positively to climate change will be an essential feature of ethical responses to such changes. The ethical objectives do not necessarily need to be pursued by policies and regulations, but by voluntary commitment based on certain principles (faith/belief) unilaterally, without aspiring for any instruction or incentive. Further, climate change may also impact population dynamics including their reproductive pattern, cultural heritage, traditional ways of living, and style of living of people through sudden submergence/reemergence of inhabited land, demographic migration, and instability of political and economic order. In this direction, brief discussion on regional collaboration, economic integration, participatory governance, and role of local administration is made below.

Regional Collaboration: South Asia with 1.7 billion people is a huge market. Consolidation of this market, in terms of consumption of goods created locally and regionally is required. The analyses of Asian Development Bank (ADB, 2014) and the World Bank (2014) cautioned South Asia's overreliance on exports, and highlighted the need to rebalance sources of growth, putting more emphasis on domestic demand. Such growth would encourage greater intraregional trade in services, and would eventually attract more local and foreign investment and enhance employment. Such an economic power may force reform in various political and economic decision making institutions even in the world.

Moreover, the eight nations may synergize their financial and infrastructural resources under SAARC umbrella. The NACP (Noosa Climate Action Plan) is an example of collaborative climate governance (Piran and

Dedekorkut-Howes, 2015). For this, the ground realities in South Asia must be recognized and appreciated correctly. The first ground reality is no country can modify the boundary with neighbor all alone. The peaceful exchange of Chhitmahals (small tiny disputed islands remained in limbo since 1948) between India and Bangladesh probably shows the way. The international border between India and Bangladesh was redrawn peacefully on July 31, 2015. The same theory could be applied in other unresolved boundary issues in South Asia (Pakistan–Afghanistan, Pakistan–India, Bangladesh–Myanmar borders, etc.). Creating a peaceful trustworthy atmosphere, hence, is essential to discuss disputed issues.

Economic Integration: South Asian economy has the potential to boost the region's economic resilience and also ensure that it contributes to global economic realignment, only if it integrates. Such integration will render the South Asia strength, stability, and predictability. The philosophy is to cooperate to prosper, and prosper to contribute to a more inclusive and environmentally sustainable brighter future for all (ADB, 2010). The possibility of such collaboration finding success is high, and lies in the fact that the borders of these countries are contiguous, and largely, and traditionally they have similar customs, values, ethics, and perception (Agranoff and McGuire, 2003; Carlson, 2007). These countries also found fighting together in recent days to thwart the effort by few Islamic fanatics who wish to systematically remove plurality from the society.

The fact is that South Asia, home to 16% of humanity and one-fourth of the world's poor, represent the world's best hope for achieving the United Nations Millennium Development Goals (MDGs; Mirza, 2007). Instead of pursuing a policy with tunnel-vision, the need for South Asia is to undertake a broad-scale approach involving all sections of the economy and government to reach to a sustainable, low-emissions development trajectory (Howes and Wyrwoll, 2012).

Participatory Governance: In governing or implementing mitigation and adaptation measures against climate change, participatory governance must play an important role. Such action would bridge the gap between policy in paper and its implementation. No wonder this aspect has been given due importance in BTG doctrine of ethical governance. The participatory climate governance may encompass Green-Growth-Engine (GGE), Public-Private Partnership (PPP), and Multilateralism. An optimum combination of all these is emerging as perfect approach for South Asia.

The *Green-Growth-Engine* (GGE) requires three assurances—a supportive economic incentive (financially viable) framework, availability of adequate physical infrastructure, and clarity in mandate (UNESCAP, 2012). While elements like privatization, deregulation, and liberalization are required to make GGE governance more efficient (Haque, 2001); absence of ethics, values, and humanity may be counterproductive (Manzoor and Ramay, 2013).

The Public-Private Partnership (PPP) is another tool of global governance that can offer both effective and legitimate governance (Backstrand, 2008). This approach brings public and private stakeholders (i.e., state and nonstate organizations) together in collective forums. The public agencies will engage in consensus-oriented decision making and regulate its implementation, while private players focus on exact and effective implementation on the ground. Although considered by some as an expensive and inefficient way of financing infrastructure (Aizava, 2017; Hall, 2015), the PPP has proved much reliable than a wholesomely private governance.

Based on a host of theoretical, empirical, and experimental evidences, Cole (2015) however concluded that Multilateralism, instead of GGE or PPP, is better to deal with climate mitigation. It involves a polycentric approach to climate policy framing and implementation than mono-centric GGE or PPP, because the former provides more opportunities for experimentation and learning to improve policies over time, and help build the mutual trust needed for increased cooperation among individuals and nations.

No governance is complete without rendering social protection to poor and vulnerable populace, and supporting inclusive growth. For example, social assistance payments even conditional (against sending children to school), guaranteed work programs, contributory pensions, old age pensions, and post-retirement incomes could help recipients withstand the social impacts of economic crises, serious illness, and natural disasters. In fact, such protection, will stimulate domestic consumption, and develop healthier, more educated, and better skilled workers (ADB, 2010). Actually, India has taken significant steps to build up her social protection systems through various schemes—Mahatma Gandhi National Rural Employment Guarantee Scheme, Atal Pension Yojna, Prime Minister Jan-Dhan Yojna, Deen Dayal Upadhya Swasthya Bima Yojna, and so on. The other South Asian countries could follow this trend to expand coverage to those employed in the informal sectors.

Local Administration: Climate governance can only be successful when there exists a fully participatory local administration (LA) that ensures incorporating all stakeholders into consensus-oriented decision-making (Ansell and Gash, 2008) to obtain a collective goal that is unachievable by any single entity alone (Silvia, 2011). However, the development in many South Asian countries suggests that municipalities/local authorities do not fully exploit their authoritative powers.

The LA must be well-informed, responsible, responsive, well-trained, and efficient to offer a wholesome balanced solution (Chatterjee et al., 2015; Cruz et al., 2007; Sivakumar and Stefanski, 2011). The LA may (a) formulate appropriate planning to regulate optimum land use, (b) ensure prompt supply of goods used for adaptation, relief, rehabilitation, and mitigation measures, (c) revise building and infrastructure standards to make them energy efficient and climate-proof, (d) manage forest conservation that can reduce carbon emissions, and create healthy ecosystems for livelihoods and industries, (e) regulate fisheries that can provide fallback options during periods of drought or shortfalls in food production, and (f) collect fiscal revenues, in the form of taxes, fees, and charges.

A Standard Operating Procedure (SOP) must be prepared by the LA to take appropriate measures to mitigate the threat. To make this model work, however, autonomy of LA in decision making, transparency in implementation, and accountability in financial dealing will be essential. The LA will also ensure participation of local people, and inclusion of ethics, values, and customs of the local people while dealing with the policy making (Tanner et al., 2009).

6.3.3 AGM: Outer Blue Layer: Mitigation and Adaptation

Once the soul is changed and purified (Section 6.3.1), and strategic roadmap for cooperative and participative governance is strategized well (Section 6.3.2), time is now ripened to initiate specific mitigating actions on focused areas. The components of the outer blue layer are the six action areas—population control, water management, agricultural reforms, clean energy, respecting nature, and hunger management (Fig. 6.3, Table 6.3). These actions need to be initiated as quickly as possible throughout South Asia in an integrated manner.

Population Control: Population in South Asia is increasing beyond its carrying capacity (Fig. 6.4). This is a primary area where action is needed as early as possible. However, substantial progress could be made by the respective governments in South Asia by undertaking the following measures (Table 6.4). Some of these measures might look like intrusion in private life of an individual, but when the entire nation is going to be impacted by a private decision, the state may certainly advice the individual appropriately.

Government of SAC must make family planning mandatory and couple may be counseled and advised to not have more than two kids. Couples could also preferably have one of the two children adopted from an orphanage home of their choice of sex different than their other child. This way the couple can ensure to have children of both sexes within the family. Government could make the process for adoption easy, friendly, quick, and respectful to complete the whole process within 90 days. Government may also offer incentives in the form of Good Citizen Cash Allowance, Life time health/family insurance and Education allowance for children for those couple who have adopted a child.

The governments of SAC must also ensure that people have easy and cheap access to contraception tools to avoid cases of unwanted pregnancies and births. All

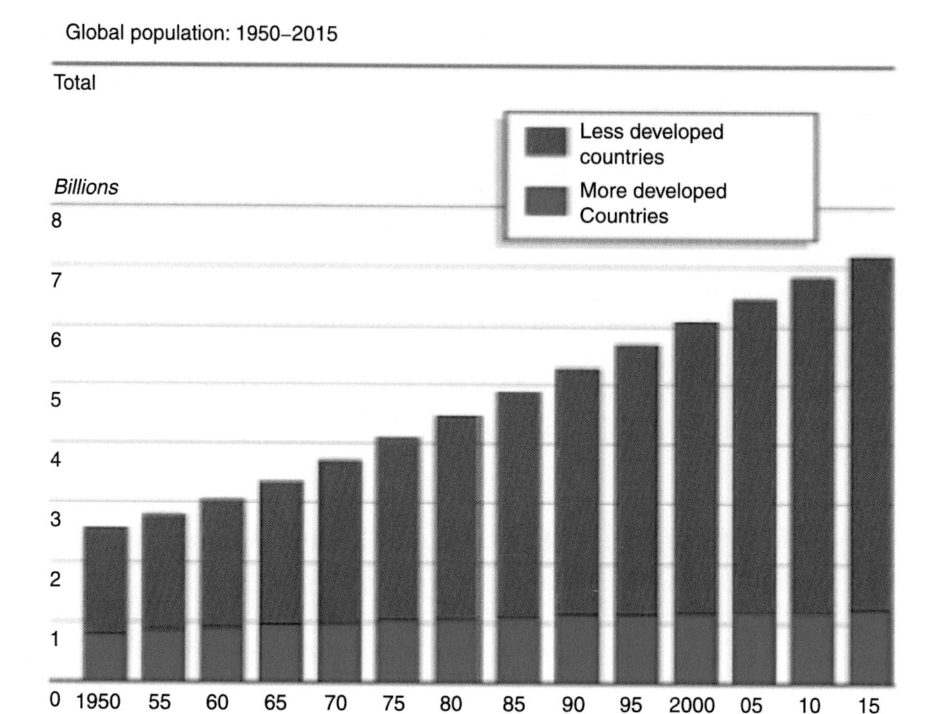

FIGURE 6.4 The growth in world population.
Note the rate and volume of increase in population is much higher in less developed countries (*red*) than that at the more the developed countries. For example, while there is only marginal increase in population in the developed world (<1 billion in 1950 to about 1.2 billion in 2015), the same in developing world increased from about 1.8 billion in 1950 to about 6 billion in 2015. Modified after *(US Bureau of the Census, 2016))*

hospitals (both private and public) may offer cheap, hygienic, and efficient birth control medicines or surgeries, as poor people have neither the means not awareness to use contraception. Use of condoms and contraceptives must be advertised and promoted along with ensuring cheap and ready access to these. Contraceptives do not only prove to be an important population control measure but also prevents spreading of sexually transmitted diseases like AIDS. Legalize abortion of unwanted pregnancies and involve NGOs to campaign for awareness and education to remove social stigma attached to sex, contraceptives, and abortion.

Water Management: Scarcity of water for human consumption and for the agricultural irrigation is posing a severe problem, not only for South Asia, but also for the entire world. The per-capita availability of water in India has declined considerably from 5177 m^3 in 1951 to only 1820 m^3 in 2001, and expected to decrease further to 1341 m^3 by 2025 and 1140 m^3 in 2050. Likewise, Pakistan has moved from water affluent country (5000 m^3 in 1947) to water stressed country (1000 m^3 in 2015), and only 800 m^3 in 2025 (SAARC

Table 6.4 Alternate governance model: prescribed strategy under BTG Doctrine

It is a fact..no more a myth

Climate is changing subtly, but surely. But may be at a rate slower than the rate argued by IPCC. Also the exact contribution percentage of anthropogenic activity is not very clear.

Bare facts

Do-not-do-anything will not solve the problem. Competitive creation of this GHG mess by the western world (Europe and North America) during 18th, 19th, and mid-20th centuries cannot be left to the developing world to clean up. Economic socialism has to come to save this Earth, while guaranteeing the humanity a decent living. The costs of reducing carbon emissions must be across the globe and over generations.

Spiritual fulcrum

People may adopt BTG doctrine to move slowly from a consumer mindset to conserving mindset. It is time to give back to the society, instead of taking. Such move could only sustain the Earth and humanity. Similar shifts have occurred earlier also, from whale oil to petroleum, from horse to automobile, from lamp light to electrification from firewood oven to cooking gas cylinders, from brick-mortar to cement-concrete, from arrow to missile, from typewriter to computer, from telephone land lines to cell phones, from letters to email/internet, and from Test cricket to 1-day to T-20.

Lifetime opportunity

GWCC (global warming and climate change) must be seen as best instrument to bring in economic socialism and may be responsible for distributing wealth of the developed countries among the poor nations legally. Fighting against a big threat can make the humanity disciplined and innovative. New technologies (CCMT-climate change mitigation technologies) will be needed to develop clean energy, including solar power. The CCMT will open up new type of jobs in manufacturing and service sector scenario. We may in fact jump into generation of progress. Similarly, encourage innovation to bring about major improvements in the areas of infrastructure, transportation, agriculture, health, living style and urban development.

Strategy of implementation

- **Population control:** Stop early marriages, encourage adoption for the second kid, offer incentive, ensure easy and cheap access to contraception tools, decriminalize abortion, Break the vicious circle: Population ↔ poverty ↔ illiteracy ↔ diseases ↔ malnutrition ↔ women empowerment ↔ Population
- **Water management:** Link the rivers of entire South Asia scientifically for irrigation, power, transport, and recharging aquifers
- **Agricultural reforms:** Initiate collective farming public-individual-private level, change crop pattern, cultivate and use Genetically Modified food
- **Clean energy:** Make it a regional campaign at all levels to opt for solar and wind energy. Use biofuel and hydrogen for vehicles
- **Resource cost:** Introduce cost to use the natural resources, minerals, fisheries, etc
- **Hunger management:** Change in food habits, livestock management, meat consumption
- **Regional cooperation**, economic integration, local administration
- Campaign for **Economic Integration** among the South Asian nations, and integration of South Asian markets. Remove uncertainty from market and strengthen economic stability. Encourage building of trust among hostile neighboring nations. Focus on adaptation measures. Government must enhance social protection to bring inclusive economic growth in the society by carrying all sections of the society together.
- **Mindset change:** Change in mindset, ethics, culture, feminism, yoga-way-of-life: Make yoga as way of life—irrespective of creed, caste, nationality, religion. Live for need not for greed. Yoga offers people a sense of "harmony with self, society, and nature" and could create a "social consciousness." Time has come to adopt an alternate strategy for sustainable South Asia.

Source: Modified from Billa and Jahan (2016), *Mazar and Goraya (2015), Dixon and Gonzalez (2017).*

2014). While India and Pakistan have real scarcity of water, Bangladesh and Nepal are plagued with lack of capacity to utilize the available water resources. Water is treated in South Asia (and even among the states within a country) as a political upmanship and may cause war in the future.

To thwart this threat, the adaptation strategy for water may include expanding rainwater harvesting, improving storage, desalination, reuse and conservation techniques, and making irrigation efficient. For example, adjusting the social patterns of water use and using technologies for more efficient water management would be a more effective adaptation strategy than building higher dams, digging deeper boreholes or setting up of highly energy intensive desalination plants. This strategy would require framing of integrated water resource and hazard management policy for the entire South Asia, and close coordination with other sectors. Recharging aquifers and linking of the rivers of the region could be few such effective steps in addressing the issue of water scarcity.

South Asia has seen real boom in groundwater irrigation over the last 35 years. Because small pumps and boreholes have proved one of the most potent land-augmenting technologies, smallholders in India, Bangladesh, Nepal Terai, and Pakistan have taken to borewell irrigation with great enthusiasm (Shah, 2007). Even if South Asia comes up with an unambiguous regulation for groundwater withdrawal the problem will not solve. What probably is needed is to have a hydrocentric vision and embrace a broader strategic view of groundwater governance (McKinsey, 2009; Wilson, 2011).

The first job under the scheme will be to identify natural aquifers and also to set aside land where the soil conditions and geology are favorable for water to infiltrate very quickly into the ground. These aquifers may be identified preferably in upstream areas of major rivers to capture and store flood water in natural underground aquifers, and then pump it out during dry spells for farmers to irrigate their land. This technology will be particularly useful for northwest South Asia that includes Afghanistan, northeast Pakistan, and northwest India.

If the South Asian political leadership really wish to fulfill the water demand and get rid from the recurring phenomenon of floods, drought, and famine then the region needs to change her past attitude, negotiate in such a way that create a win-win situation to all riparian neighbors. For all neighbors to get their fair share and benefits equitably, linking of rivers in South Asia is a very viable project and may be expeditiously implemented in phases (Mukhopadhyay, 2008). Hence, discussions and agreements on river water sharing must move from being bilateral to multilateral. Comanagement of transboundary rivers for mutual benefit, especially toward generation and sharing of power, drinking, irrigation, and transport are extremely important (Merrey et al., 2017). In fact sharing of river water could bring peace and amity in this region.

The leaders of South Asian nations must show statesmanship to frame a concerted policy and mechanism that go beyond national boundaries. To start with, all the rivers of this region must be mapped in detail with regard to catchment area, flow rate in different seasons, and gradient. Interlinking of rivers and water bodies of the entire South Asia could brought about major changes to (i) help reduce floods during the monsoon by flowing excess waters to other places, (ii) avoid drought by supplying excess waters from other areas, (iii) initiate least expensive transport of goods from one place to other, (iv) help generate hydroelectricity for the entire South Asia grid, and (v) help increase greenery all around and aforestation helping to the cause of carbon sequestration.

Agricultural Reforms: As the world's population expected to grow from now 7 billion (700 crores) to 9 billion (900 crores) by 2050, pressure on agriculture will increase (Fig. 6.4). However, production of wheat—one of the world's most important staple crops—is set to fall by 6% for every 1 °C rise in temperature (Asseng et al., 2014). Hence, the adaptive strategy could include—(a) modifying pattern, variety, place, and date/time of crop sowing; (b) introducing cooperative farming with several square kilometers of land being cultivated together in the form of industry; and (c) encouraging genetically modified (GM) crop. In fact, about 54% of methane, 80% of nitrous oxide, and almost 100% of terrestrial carbon dioxide are contributed by agriculture alone (see Chapter 3). Hence, there is an urgent need to change the present agricultural practices to include collective farming, changing crop pattern and sowing specific GM crop.

Agriculture should be given the status of an industry. The contiguous cultivable land of a reasonably large area (several km^2) could be "leased" (*not sold*) to any Public-Private Consortium (PPC) or to Private Enterprise (PE) after payment of existing/agreeable market price to the land-owning farmers to form Agricultural farms. Farming by PPC/PE over such a vast agricultural area in a scientific manner with latest technology and collective dispensation would be both economically and commercially viable. Besides, one able-bodied person from each family of whose land has been leased to PPC/PE may be given a job to work in the agricultural farm. Sowing, cultivation, storing, and marketing of the crop in the local or global market will be the responsibilities of concerned PPC/PE. Farmers will receive salary at the end of every month and thus could be insulated from crop loss or financial bankruptcy or any other eventuality in future.

Government will only facilitate the entire process and could act as a distant-regulatory body to provide level-playing field to stakeholders and neutralize exploitation of farmers, if any. This mechanism will offer financial stability to farmers, stop them from committing suicides, while the land will continue to be theirs and being effectively used to benefit the humankind. The fact is that the land deserves to be used more scientifically and rationally rather than keeping hostage to rituals, customs, and hollow emotion.

With change in climate, the farmers of South Asia must also change their historic mindset to food and agriculture. The people must accept that specific agricultural fields are meant for some specific types of crop. The Farmer Awareness Program being conducted by Indian Meteorological Department and also by BOBLME (Bay of Bengal Large Marine Ecosystem) in this direction may be copied by other South Asian nations with appropriate ground level modification, as needed. Hence, a structured, focused countrywide testing of the soil, climate, weather, rainfall, and availability of ground water are required to be undertaken first, to identify and earmark each agricultural farm with the most suitable crop.

The Genetically Modified (GM) crop are those seeds, whose DNA has been modified through genetic engineering to introduce a new trait to an existing naturally produced crop. Such crops would have increased resistance to pests, diseases, and spoilage, but with increased nutritional value and improved shelf life. To feed the increasing population, enhanced demand for land for infrastructural purposes, and increasing intrusion of saline water in the coastal landspaces, the overall shortage of land in South Asian countries has been quite serious. Moreover, the agricultural yield is predicted to reduce by 10%–35% by 2050 because of climate change, and the South Asia must make maximum use of whatever land is available. In this regard, cultivation of GM crop is one option through which good healthy and substantial amount of food grains could be produced most optimally in the land made available. The GM crops are sustainable, comparatively cheap, nutritious, and probably can take care of increasing mouths in the South Asia, as well as in the world. However, greater research is needed on GM crop to make it happen and remove the doubts from the minds of South Asian population. In fact, South Asian countries may start investing in genetic mapping of climatically adaptable food items, forest tree species, and to foster economic resiliency.

The GM crop being produced presently includes apples, potato, tomato, soyabeans, papaya, cotton, corn, rapeseed, sugar beet, sugarcane, melon, rose, argentine canola, carnation, petunia, squash, chicory, plum, polish canola, sweet pepper, creeping bent-grass, poplar, tobacco, flaxseed, linseed, maize, rice, and wheat. Cultivating GM would mean higher crop and livestock yields coupled with lower pesticide and fertilizer applications, less demanding production techniques, higher product quality, better storage and easier processing, or enhanced methods to monitor the health of crop and yield. India has presently more than 9 million hectares under GM cultivation, followed by Pakistan (1–3 million hectares).

Clean Energy: Details of high priority innovations in climate technology and the emergence of highly potential climate industry are discussed in detail in Chapter 5. Both innovations and industry needs investment of very high amount in terms of research and infrastructure. The clean energy production is cost-intensive proposition and hence requires collaboration, partnership or cooperation among various nations in South Asia. Such investment could come both from public and private companies. To make South Asia take lead in developing climate technology it would be required to confront perceived impediments, invest money in renewable electricity grid for the entire region, and obtain the moral, financial, and regulatory support from the respective government.

Resource Formation Cost (RFC): Tagore once said "The roots below the Earth claim no rewards for making the branches fruitful." Following this philosophy, it is time South Asian governments introduce RFC. In fact nobody pays for the resources they are using—fishermen do not pay for fishes the nature is creating, miners not for the ores, and farmers not for the soil. But they spend money downstream on the fishing/mining/farming process, and on the mechanism to extract or refine these resources. These entire public and private entrepreneurs take the mother Earth that produces the resources, granted. The painstaking efforts of nature for several millions of years go unrewarded. Once this RFC is introduced, hopefully the reckless extraction of resources and its wastages will come down. The RFC could be a small measure to balance between the natural resource supply and the human demand on the environment to promote sustainability (Biggs et al., 2015). In South Asia this balance often breaks (Rasul, 2014). A quick look on the problems faced in coastal zone, and with health and tourism is given later.

The best way to deal with coastal erosion, inundation, loss of land, and coastal pollution is to go for soft adaptation using natural mechanism, such as dune reinforcement, creation and management of wetlands, and protection of existing natural barriers. It is prudent to avoid seawall, dikes, and hard structures and avoid short-term solutions that may result in long-term problems. Government may strictly implement a rational geomorphology and biodiversity specific CRZ regulation after discontinuing the existing blanket ban on development irrespective of type of coast. This, coupled with effective land-use policy, building codes and insurance would help restore mangrove systems to protect coastlines. It would also be advisable to adapt a flexible infrastructure and habitable pattern along the coastal zone.

Health is an important component of the prescribed ethical governance model. The higher organic: black carbon (>5, except in cities) seen over South Asian region indicate the significant contribution of biomass burning (Babu et al., 2011; Ramanathan and Feng, 2009). It is suggested to initiate surveillance measures for climate-sensitive diseases in vulnerable or high-risk areas. Setting up of databases and early warning systems on climate-sensitive vector- and water-borne diseases, and to track their geographic distribution will be of much help. Government must prioritize research and education on climate-related diseases, and train health professionals and educate communities.

Compared to Sri Lanka, Bangladesh, and Nepal, India's progress however toward the achievement of its Millennium Development Goals (MDG) has been quite dismal. Despite having their own "local" problems, Bangladesh and Nepal have achieved or nearly achieved many of the MDG targets of optimal maternal and child health and nutrition (Fidler, 2010). Sri Lanka is already in its post-MDG phase. Poor governance, lack of political will, divergence of effort, and the lack of a transparent dedicated prochild and promother health system could be the reasons for such poor performance in India and rest of South Asia (Rajan et al., 2014).

Similarly, special attention may be given to create a new taste to entertainment and tourism industries by diversifying attraction. Adventure sports, for example, may be restricted to skiing and nature's walk on glaciers, paragliding, and ropeway, instead of vehicle traversing the sensitive mountainous and coastal ecosystem. The plan must include estimating carrying capacity of any tourism industry/place and ensuring integrated linkage to other sectors. Improvement of the road network could spur new livelihood opportunities and boost sustainable growth.

Hunger Management: Nearly one out of every seven people on Earth now suffers from chronic hunger or food insecurity. Climate change could further negatively impact food production causing widespread hunger. In addition to fossil fuel, livestock contributes about 15% of GHG emission (=~7 billion tons of CO_2) through consumption of meat and milk. This contribution comes from methane derived from enteric fermentation in animal's digestive system (39%), nitrogen oxides through fertilizers used to grow feed for livestock (21%), and from the manure produced by these animals (26%). The rest 14% of GHG concentration was caused by transportation, processing, and deforestation to provide land for growing feed (TOI, 2015).

It is found that among all meats, beef has the highest carbon emitting potential. In fact, a kilogram of beef emits 22.6 kg CO_2 (equivalent to running a car for 160 km), compared to 2.5 kg from pork, 1.6 from poultry, and 1.3 kg from milk. The consumption of beef worldwide has increased from 70 MT in 1960s to 278 MT in 2009 and is projected to be about 460 MT by 2050. Compared to meat, plant cultivation is responsible for much less emissions: for example, in contrast to Beef, one kg of wheat was found to emit just 0.8 kg CO_2. The average daily per person meat consumption is little more than 300 g in USA, Australia, and New Zealand; about 200 g in Europe, Brazil, Argentina, and Venezuela; about 160 g in China, and only 12 g in India. Hence, reducing consumption of meat could be a great battle win over climate change (TOI, 2015).

6.4 EPILOGUE

This book writing was taken up after recognizing that the global warming and climate change (GWCC) is caused both by natural and anthropogenic activities, and appreciating that the adverse impact of GWCC on resources like food, water, agriculture, human-health, ecology, and environment of South Asia will only increase in years to come. The other reasons to write this book were to underline the need to understand the threats of GWCC on South Asia vis-à-vis the global carbon politics, and accepting that South Asia needs an innovative participatory public policy that must address economic and social disparity, and augur inclusive growth in the society. This essay desires that all the nations in South Asia must cooperate, collaborate, and build trust based on ethics, values, and egalitarianism to enable sustainable, balanced, and purposeful growth in the region. It also comprehends that time has come to change our mindset to transform the society from greed-based to need-based.

Governance means providing the necessary policy and legislative framework to promote and protect the mental, physical, and financial health of a population (health, education and security). And if anything failed in South Asia miserably over the last 100 years, it has been the governance (Mukhopadhyay, 2018; Peterson et al., 2017). Inadequate and inappropriate financing and rampant corruption emerged as the biggest challenges for good governance. In addition, South Asian countries have not timed these reforms properly and the slow pace of economic transformation in this region is proving disastrous (Rana and Chia, 2015). The first round of reforms in South Asia in early nineties was successful that focused on macroeconomic fiscal reforms. This reform unshackled government controls leading to private sector driven economic growth and job creation. However absence of second generation reforms in terms of increasing transparency, improving law and order and reforming the legislation and judiciary have kept this region way behind other regions of the world.

It's a great relief and heartening to note that all the eight South Asian countries have chalked out their adaptation and mitigation strategies, quite comprehensively (see Chapter 4). However, an element of disconnection between policies and people is very much distinct. Consequently, the concept of transforming climate change threats into opportunities, such as job creation, is only slowly emerging in South Asia. Such campaign however could pick up with increase in mutual trust, cooperation among the countries in South Asia, and with integration of markets and economies of the region.

As mentioned earlier in mitigating the climate change impact, if public administration is considered as the hardware, then mindset becomes software. To make public administration successful, visible, and relevant, the mindset of the individual and society needs moderation. The BTG doctrine (*Buddha-Tagore-Gandhi*) in this regard is introduced here which links our actions to our ethical minds. Leaving behind the paths of capital economy (emphasizes on profit and aims for perpetual growth) and religious economy (divides the world into narrow religious bigots/fraternity), the world must follow the unbeaten track of ethical economy that provides a value-based spiritually inspired balanced standard of living in complete harmony with the biosphere, hydrosphere, and atmosphere.

The BTG doctrine discussed in Section 6.3 (Tables 6.3, 6.4), (Fig. 6.3) as an alternate governance model for South Asia relies heavily on ethics, repairing in the process ruptured relation between soul and body, between environment and economy, and similarly between human and nature. The ethical teachings of Lord Buddha, Poet Tagore, and philosopher Gandhi for a society devoid of greed, manipulations, deception, and shortcuts could form the base of the alternate governance. While soul must seed the bonding between human and environment, the mind should take forward the goodness of the society to the neighbors (generating trust, confidence) to ultimately work together for integrated development of South Asia.

However ethical governance is no replacement for the major bottleneck in South Asia, which has been negative bureaucracy, corruption and the valley of distrust among the nations (Table 6.5). For example average number of

Table 6.5 Governance factors in South Asia: investment climate

Parameters	AFG	BNG	BHU	IND	NEP	PAK	SRL
Access to finance	49	23	19	15	40	22	33
Political environment	76	76	12	16	85	34	13
Crime	58	8	1	5	14	35	7
Taxes	56	20	24	31	23	55	41
Corruption	62	49	4	36	42	64	15
Informality	33	9	10	17	29	12	28
Infrastructure	81	55	29	26	79	79	36
Electricity	66	52	14	21	69	75	26
Telecom	59	3	15	4	3	14	6
Transport	43	15	14	10	32	27	12
Labor regulations	11	3	15	11	3	12	13
Work force education	53	16	14	9	9	23	16
Trade and customs	47	8	9	12	29	30	31
GNI per capita ($)	690	1010	2330	157	730	136	3170
GDP growth (annual)	1.9	6	2	6.9	3.8	4.4	7.3
Population (million)	30.6	156.6	0.8	1252.1	27.8	182.01	20.5
Voice and accountability	-	33	44	61	33	27	28
Government effectiveness	-	22	63	45	20	22	57
Regulatory quality	-	18	15	35	21	28	50
Rule of law	-	26	68	54	28	24	52

AFG, *Afghanistan;* BHU, *Bhutan;* BNG, *Bangladesh;* CHN, *China;* GDP, *Gross Domestic Product,* GNI, *Gross National Income per capita (2015);* IND, *India; MAL, Maldives;* NEP, *Nepal;* PAK, *Pakistan,* SRL, *Sri Lanka;*
Sources: Modified from *Billa and Jahan (2016) Mazar and Goraya (2015), Dixon and Gonzalez (2017)*

documents and days needed to export or import goods by countries in South Asia and ASEAN are notable. It seem to take 8.5 and 9.0 documents to export or import goods from and to South Asia compared to 6.7 and 7.8 in ASEAN countries, respectively. Similarly, it takes 31.5 and 31.1 days for South Asian countries to export or import goods as compared to 22 and 21.7 days for ASEAN countries (World Bank, 2014).

There is no denying of the fact, however, that since the time this subcontinent obtained freedom by the middle of the 20th century, it has progressed a lot. However, the three most important aspects that still plague this region are illiteracy, poverty, and corruption. South Asia is a clear case of unrealized potential. The partisan politics, structural impediments, and extreme greediness of few people are holding this region to ransom. In addition to these, the religious violence, fanatic-extremism, and border disputes have created gulf of distrust among the nations.

In this burgeoning scenario, formulating a policy to mitigate the impact of climate change for the entire South Asia is a challenging task. Social justice cannot come through job reservation (a misplaced priority of the governments), but probably by removing illiteracy and bringing in quality education. Lack of education germinates inefficient governance, low industrial productivity, and institutional mediocrity. Running airlines or roadways cannot be the job of a government, but these are happening in all South Asian countries. Again, state and religion should be kept separated, as this is another area to create avoidable communal tension in the society. It is difficult to believe that a region which could land her spacecraft on Moon and Mars could not deal with infant mortality (healthcare), or remove infrastructure inadequacy, or deal with hunger and poverty.

The BTG Doctrine sees a vision for South Asia in 2035 (if its provisions are followed optimally), where people would live under the jurisdiction of ethical governance powered by circular economy where nothing is wasted. Around the year 2035, the "cause of nature" would be held at the highest esteem—valued and respected. The basic understanding of BTG is that people cannot live without nature, and without people there is no economy. Making peace within self is prerequisite before making peace with nature. Yoga could play a major role in this regard. All components of nature are not only interlinked, but interdependent also. BTG hints for meaningful time-bound trust-worthy dialogue between parties on technical issues without being emotional, and personal. The BTG Doctrine could help bring much needed sustainability and respectability in man–nature relationship, and take the South Asia to a new height of development. Even if only 60% of the provisions displayed in three layers of BTG wheel (Fig. 6.4, Tables 6.3, 6.4) are integrated optimally, a visible change could be witnessed in South Asia. A roadmap to implement new measures and mindset is given in Table 6.6.

Table 6.6 Possible future strategy for South Asia till 2035

01 Regional collaboration • Settle border demarcation based on ground realities. Avoid hollow emotion and dreams • Encourage people to people contact • Open borders for artists/students/scientists/sportspersons • Open 500 m wide rail-road corridor from Central Asia to SE Asia via SA • Make South Asia, not only producer and exporter, but also a consumer of its own production	**02 Profound action** • Bring down population from 1.7 to 1.5 billions • Reduce population density from 341 to 300 • Integrate economy, explore having uniform currency • Explore having borderless trade • Reduce days and documents needed for export/import • Make solar power the order of the day • Undertake intense R&D to make best use of plastic—in road/bridge making/construction • SACs are to deradicalize hate apparatus, and remove barriers that divide people in terms of caste, creed, religion, and sect.
03 Agriculture and flood control • Link most of the rivers in phases to ensure maximum areas in South Asia under irrigation and also to minimise flood and drought • Also • Form agricultural farms • Modify crop seeding pattern • Research and use GM food to its full	**04 Health and education** • Set up low cost but top quality healthcare for all people through insurance/otherwise • Give education and farmer's cause top priority • Provide low cost low emission chulhha to all rural households • Spend 8% of GDP on education, and another 8% on health • Support primary education, skill development
05 Quality of life and lifestyle • Replace all plastics with glass and jute • Use rechargeable batteries, biodegradable sanitary pads • Improve HDI standing from 131 to 110 • Reduce per capita CO_2 emission from 1.05 to 0.94	**06 Behavior and philosophy** • Follow the teachings of Lord Buddha, Tagore and Gandhi to live for need, not for greed • Women diplomats are to increase by 40% • SACs are to evolve based on ethics, values, culture and knowledge • Bring human close to nature

CO2 emission: National CO2 emission in kilotons 2011 figure, and increased difference from 2010, Area 50,02,609 km2, Population 1,706,508,995 (2015); Modified after *UNDP (2013), HDI, Human Development Index (an assessment of long term progress in health education and income indicators, HDI Report 2014, UNDP, ranking out of 187 countries). A clear turnaround at 2030 is envisaged in terms of population, CO2 emission, HDI standing, and Energy generation.*

Another main challenge to governance in South Asia is the increasing gap between the rich and the poor. The coexistence of poverty with affluence indicates slow and inadequate response to this disparity (Jamil et al., 2013). Afghanistan is torn between war and terrorism with infrastructure deficiency, India is struggling with religion and caste tensions, and the postconflict states like Nepal (Maoist) and Sri Lanka (LTTE) are still struggling to come to terms. Pakistan is tattered between Islamic fundamentalists on one hand and Baluchistan separatist movement on the other. Bangladesh is possibly the only country in South Asia with less ethnic and religious tension, although the country is divided very sharply along political lines, especially between the two major alliances—moderate Awami League and the Islamist Bangladesh Nationalist Party.

South Asia probably needs a long-term comprehensive (not necessarily popular) policy that would address these issues of discrepancies squarely and politically in a time-bound manner. Transparent and accountable governance based on the philosophy of Democratic Socialism in its purest form may probably offer answer to many of the woes of South Asia. The new governance principle of South Asia must focus on wellness of its millions of middle- and low-income groups. The rich also must be encouraged to aspire beyond their private affluence to reach out to the poor and needy through charities.

A politically stable and ethically governed inclusive South Asia could as well be the fulcrum of global shifts in economic, political, and even sociocultural power equations (Bice and Sullivan, 2013). Much to the disappointment of admirers, South Asia today is crumbling under its own weight of nonperformance, poor governance, and widespread corruption, despite having a vast potential of the population below 35 years, abundant natural resources and a rich ancient culture. With almost no competition among government departments, efficiency is unfortunately replaced by a fine mechanism-of-managing-things (juggad/dalal/middle-man) through hooks or crooks. Added to this is the *chalta-hai* attitude reflecting complete lack of accountability (Elliot, 2015).

The unpreparedness and inconsistency of India and entire South Asia are vivid in various hues throughout the history. The gaps between what is said and what is done are phenomenal. Political parties are formed following slightest ego-clash like clubs in a countryside (muhallahs/villages) without any philosophy, aim, and objectives. Vote-bank politics throughout the region is playing havoc. Democracy is tottered, administration is indisciplined and corrupt, and judiciary is slow and orthodox. Welfare economics found its easy way in South Asia unfortunately through job reservation, and other special powers to rich and influential.

Time is running out. South Asia needs to choose between an ill-prepared democracy that continues to provide a cover for the country's inequality, injustice, corruption, ill-governance, or a new system which offers able and corruption-free governance taking advantage of history, culture, brain power, and aspiration of a billion youth. The task is to deliver economic growth, reduce inflation, take proactive decisions, streamline procedures, and develop new partnership. In summary of what needed now is to have ethically performing, honest, efficient, environmentally responsible and plural social democratic set up.

Bibliography

Achar, B.N.N., 2005. Planetarium Software and the Date of the Mahabharata War. In: Rukmani, T.S. (Ed.), The Mahabharata: what is not here is nowhere else. Munshiram Manoharlal Publishers, New Delhi, India, pp. 247–263.

ADB, 2010. Asian Development Bank, Research Agencies Target Food Security, Nutrition Gains in Asia.

ADB, 2012. Asian Development Outlook 2012. Confronting Rising Equality in Asia. ADB, Mandaluyang city, Philippines, pp. 1–298.

ADB, 2014. Asian Development Bank Annual Report 2014. ADB, Mandaluyang city, Philippines, pp. 1–52.

ADB, 2016. Asian Development Outlook 2016 Asia's Potential Growth. ADB, Mandaluyang city, Philippines, pp. 1–317.

Adger, W.N., Barnett, J., Brown, K., Marshal, N., O'Brien, K., 2013. Cultural dimensions of climate change impacts and adaptation. Nat. Clim. Change 3, 112–117.

Adger, W.N., Agarwala, S., Mirza, M.M.Q., Conde, C., O'Brien, K., Pulhin, J., Pulwart. R., Smit, B., Takahasi, K., 2007. Assessment of adaptation practices, options, constraints and capacity. Climate Change 2007 Impacts, Adaptation and Vulnerability. Contribution of Working Group II to the Fourth Assessment Report of the Intergovernmental Panel on Climate Change, Parry, M.L., Canziani, O.F., Palutikof, J.P., van der Linden, P.J., Hansen, C.E., (Eds.), Cambridge University Press, Cambridge, UK, pp. 717–743.

Adve, N., 2013. Another climate change event. The Hindu.

Afghanistan Government Portal, 2015. INDC Submission to UNFCCC From GIRoA (Government of Islamic Republic of Afghanistan), 8 pp.

Agassiz, J.L.R., 1841. On glaciers and boulders in Switzerland. Report of the British Association for 1840, pp. 113–114.

Agranoff, R., McGuire, M., 2003. Collaborative Public Management: New Strategies for Local Government. Georgetown University Press, Washington, DC.

Ahmed, M., Suphachalasai, S., 2014. Assessing the Costs of Climate Change and Adaptation in South Asia. Asian Development Bank, UK Aid, Mandaluyong City, Philippines, pp. 1–163.

Ahmed, A.U., 2006. Bangladesh climate change impacts vulnerability. Comprehensive Disaster Management Program. Department of Environment, Government of Bangladesh, pp. 1–49.

Aizava, M., 2017. Scoping Study on Public-Private Partners, 2017. FfDO working paper. Available from: http://www.un.org/esa/ffd-follow-up/inter-agency-taskforce.html.

Akhtar, S., 2017. Achieving the Sustainable Development Goals in South Asia. Key Policy Priorities and Implementation Challenges. UNESCAP (UN Economic and Social commission for Asia and the Pacific), South and South-West Asia office, C2, Qutab Institutional Area New Delhi, 110016, India, pp. 1–60.

Allison, M.A., 1998. Geologic framework and environmental status of the Ganges Brahmaputra Delta. J. Coast. Res. 14, 826–836.

Alvarez, W., 2003. Comparing the evidence relevant to impact and flood basalt at time of major extinctions. Astrobiology 3, 153–161.

Anand, S., Sen, A., 1994. Sustainable Human Development: Concepts and Priorities. Harvard University, p. 79.

Anenberg, S.C., Schwartz, J., Shindell, D., Amann, M., Faluvegi, G., Klimont, Z., Janssens-Maenhout, G., Pozzoli, L., Van Dingenen, R., Vignati, E., Emberson, L., Muller, N.Z., West, J.J., Williams, M., Demkine, V., Hicks, W.K., Kuylenstierna, J., Raes, F., Ramanathan, V., 2012. Global air quality and health co-benefits of mitigating near-term climate change through methane and black carbon emission controls. Environ. Health Perspect. 120, 831–839.

Ansell, C., Gash, A., 2008. Collaborative governance in theory and practice. J. Public Adm. Res. Theory 18 (4), 543–571.

Appell, D., 2008. Let Us Get Real on the Environment. The Guardian, London.

Archer, D., 2005. Fate of fossil fuel CO_2 in geological time scale. J. Geophys. Res. 110, 1–6, C09S05.

Archer, D., Eby, M., Brovkin, V., Ridgewell, A., Cao, L., Mikolajewicz, U., Caldeira, K., Matsumoto, K., Munhoven, G., Montenegro, A., Tokos, K., 2009. Atmospheric life time of fossil fuel CO_2. Annu. Rev. Earth Planet. Sci. 37, 117–134.

Ashworth, R., Nahle, N., Schreuder, H., 2011. Greenhouse Gases in the Atmosphere Cool the Earth! Available from: www.hampshireskeptics.org.

Asseng, S., Ewert, F., Martre, P., Rötter, R.P., Lobell, D.B., Cammarano, D., Kimball, B.A., Ottman, M.J., Wall, G.W., White, J.W., Reynolds, M.P., Alderman, P.D., Prasad, P.V.V., Aggarwal, P.K., Anothai, J., Basso, B., Biernath, C., Challinor, A.J., De Sanctis, G., Doltra, J., Fereres, E., Garcia-Vila, M., Gayler, S., Hoogenboom, G., Hunt, L.A., Izaurralde, R.C., Jabloun, M., Jones, C.D., Kersebaum, K.C., Koehler, A.K., Müller, C., Naresh Kumar, S., NendelC, O'Leary, G., Olesen, J.E., Palosuo, T., Priesack, E., Eyshi Rezaei, E., Ruane, A.C., Semenov, M.A., Shcherbak, I., Stöckle, C., Stratonovitch, P., Streck, T., Supit, I., Tao, F., Thorburn, P., Waha, K., Wang, E., Wallach, D., Wolf, J., Zhao, Z., Zhu, Y., 2015. Rising temperatures reduce global wheat production. Nat. Clim, Chang. 5, 143–147.

Audubon, 2016. Audubon's Position on Wind Power. Available from: www.audubon.org.

Awadalla, C., Coutinho-Sledge, P., Criscitiello, A., Gorecki, J., Sapra, S., 2015. Climate Change and Feminist Environmentalisms: Closing Remarks, 2015. Available from: www.thefeministwire.com.

Backstrand, K., 2008. Accountability of networked climate governance: the rise of transnational climate partnerships. Glob. Environ. Polit. 8 (3), 74–102.

Babu, S.S., Chaubey, J.P., Moorthy, K.K., Gogoi, M.M., Kompalli, S.K., Sreekanth, V., Bagare, S.P., Bhatt, B.C., Gaur, V.K., Prabhu, T.P., Singh, N.S., 2011. High altitude (~4520 m amsl) measurements of black carbon aerosols over western trans-Himalayas: seasonal heterogeneity and source apportionment. J. Geophys. Res. 116, D24201. doi: 10.1029/2011JD016722.

Bajracharya, S.R., Mool, P.K., Shreshta, B.R., 2008. Global climate change and melting of Himalayan glaciers. In: Ranade, P.S. (Ed.), Melting glaciers and rising sea levels : impacts and implications. The ICFAI University press, India, pp. 28–46.

Bala, G., 2011. Should we do geo-engineering research? CAOS and Divecha Center for Climate Change. Indian Institute of Science, Bangalore, pp. 1–31.

Bäthge, S., 2010. Climate Change and Gender: Economic Empowerment of Women Through Climate Mitigation and Adaptation? Working Paper, The Governance Cluster. Deutsche Gesellschaft für Technische Zusammenarbeit (GTZ) GmbH, Postfach 5180–65726 Eschborn Germany, pp. 1–22.

Barnosky, A.D., 2011. Earth on the brink of extinction? Hindustan Times.

Benchaita, T., 2013. Greenhouse gas emissions from new petrochemical plants. Background Information Paper for the Elaboration of Technical Notes and Guidelines for IDB Pojects. Technical Note No. IDB-TN-562. Inter-American Development Bank, pp. 1–84.

Biggs, E.M., Bruce, E., Boruff, B., Duncan, J.M.A., Horsley, J., Pauli, N., McNeill, K., Neef, A., Van Ogtrop, F., Curnow, J., Haworth, B., Duce, S., Imanari, Y., 2015. Sustainable development and the water-energy-food nexus: A perspective on livelihoods. Environ. Sci. Policy 54, 389–397.

Biomass Energy for Rural India, Carbon mitigation Report (BERI Project), 2010. Funded by UNDP (United Nations Development Program) New Delhi, p. 84.

Bewick, R., Sanchez, J.P., McInnes, C.R., 2012. Gravitationally bound geoengineering dust shade at the inner Lagrange point. Adv. Space Res. 50 (10), 1405–1410.

Bhatta, G.D., Aggarwal, P.K., Kristjanson, P., Shrivastava, A.K., 2016. Climatic and non-climatic factors influencing changing agricultural practices across different rainfall regimes in South Asia. Curr. Sci. 110, 1272–1281.

Bhattacharjee, A., 2011. Modern management through ancient Indian wisdom: towards a more sustainable paradigm. School Manage. Sci. Bull. Varanasi 4(1), 1–24.

Bhattacharjee, A., Maitra, A., 2011. Evaluation and Prioritization of Parameters Required to Design Vulnerability Index for Deltaic Regions (M.Sc. Dissertation) submitted to National Inst Oceanography, Dona Paula, Goa.

Bhattacharya, A.K., 2008. The morphodynamic setting and substrate behavior of the Sundarbans mangrove wetland of India. Sarovar Saurabh 4, 2–9.

Bhattacharya, P., Chatterjee, D., Jacks, G., 1997. Occurrence of Arsenic contaminated ground water in alluvial aquifers from delta plains, eastern India: Options for safe drinking water supply. Int. J. Water Resour. Dev. 13, 79–92.

Bhattacharyya, A., Werz, M., 2012. Climate Change, Migration, and Conflict in South Asia. Rising Tensions and Policy Options across the Subcontinent, Center for American Progress, Heinrich Böll Stiffung. Available from: www.Americanprogress.org.

Bhoi, R., Ali, S.M., 2014. Potential hydropower plant in India and its impact on environment. Int. J. Eng. Trends Technol. 10 (3), 1–6.

Bice, S., Sullivan, H., 2013. Towards a research agenda for public policy in the Asian century. 1st International Conference on Public Policy, pp. 1–18.

Bierbaum, R., Fay, M., 2010. World Development Report 2010: Development and Climate Change. World Bank, Washington, DC.

Billa, A.H.M.M., Jahan, N., 2016. Challenges and prospects of aid agencies models of governance in South Asia. World J. Soc. Sci. 16 (2), 155–166.

http://www.blewbury.co.uk/

Boas, F., 1928. Anthropology and Modern Life. Dover Publications, New York, NY.

Boesche, R., 2003. Kautilya's Arthasastra on war and diplomacy in ancient India. J. Milit. Hist. 67 (1), 9–37.

Bond, D.P.G., Wignall, P.B., 2014. Large igneous provinces and mass extinctions. In: Keller, G., Kerr, A.C. (Eds.), Volcanism, Impacts, and Mass Extinctions: Causes and Effects. Geological Society of America Special Paper 2014, 505, pp. 29–55.

Boyd, P.W., Jickells, T., Law, C.S., Blain, S., Boyle, E.A., Buesseler, K.O., Coale, K.H., Cullen, J.J., de Baar, H.J.W., Follows, M., Harvey, M., Lancelot, C., Levasseur, M., Owens, N.P.J., Pollard, R., Rivkin, R.B., Sarmiento, J., Schoemann, V., Smetacek, V., Takeda, S., Tsuda, A., Turner, S., Watson, A.J., 2007. Mesoscale iron enrichment experiments 1993-2005: synthesis and future directions. Science 315 (5812), 612–617.

Bozin, S.E., Goodyear, G.C., 1968. Growth of ionization currents in carbon tetrafluoride and hexafluoroethane. J. Phys. D Appl. Phys. 1, 327–334.

Brasseur, G.P., Garnier, C., 2013. Mitigation, adaptation, or climate engineering? Theor. Inq. Law 14 (1), 1–20.

Brohan, P., Kennedy, J.J., Harris, I., Tett, S.F.B., Jones, P.D., 2006. Uncertainty estimates in regional global observed temperatures: a new data set from 1850. J. Geophys. Res. 111, 1–21, D12106.

Bryson, R.A., 1997. The paradigm of climatology: an essay. Bull. Am. Meteorol. Soc. 78, 450–456.

Burnell, P., 2012. Democracy, democratization and climate change: complex relationships. Democretization 19 (5), 813–842.

Burtman, V.S., Molnar, P., 1993. Geological and geophysical evidence for deep subduction of continental crust beneath the Pamir, US Geological Survey Special Paper 281.

Butler, J.H., 2013. CO_2 at NOAA's Mauna Loa Observatory reaches new milestone: tops 400 ppm. Global Monitoring Division—ESRL-GMD.

Byravan, S., Rajan, S.C., 2012. An evaluation of India's national action plan on climate change IFMR and Humanities and Social Sciences, IIT Madras, Centre for Development Finance (CDF), pp. 1–32. Available from: www.indiaclimatemissions.org.

Caldeira, K., Bala, G., Cao, L., 2013. The science of geoengineering. Annu. Rev. Earth Planet. Sci. 41, 231–256.

Caldeira, K., Wickett, M.E., 2003. Oceanography: anthropogenic carbon and ocean pH. Nature 425, 365.

Carlson, C., 2007. A Practical Guide to Collaborative Governance. Policy Consensus Initiative, Portland, OR.

Casillas, C.E., Kammen, D.M., 2010. The energy-poverty-climate nexus [Policy Forum]. Science 330, 1181–1182.

Census of India, 2011. Ministry of Home Affairs, Government of India, India.

CEC, 2017. Climate Engineering Conference 2017. Institute for Advanced Sustainability Studies, Germany.

Chakraborty, P., Vudamala, K., Chennuri, K., Armoury, K., Linsy, P., Ramteke, D., Sebastian, T., Jayachandran, S., Naik, C., Naik, R., Nath, B.N., 2016. Mercury profiles in sediment from the marginal high of Arabian Sea: an indicator of increasing anthropogenic Hg input. Environ. Sci. Pollut. Res. 23 (9), 8529–8538.

Chandrasekharam, D., Chandrasekhar, V., 2015. Geothermal resource India, country update. In: Proceedings World Geothermal Congress. Melbourne, Australia, pp. 1–8.

Chatterjee, N., 2012. Vulnerability Assessment of the Indian Sundarbans (M.Sc. Dissertation) submitted to National Intitute of Oceanography, Donapaual, Goa, India. , 95 p.

Chatterjee, S., Goswami, A., Scotese, C.R., 2013. The longest voyage: tectonic, magmatic, and paleoclimatic evolution of the Indian plate during its northward flight from Gondwana to Asia. Gondwana Res. 23, 238–267.

Chatterjee, N., Mukhopadhyay, R., Mitra, D., 2015. Decadal changes in shoreline pattern in Sundarbans. J. Coast. Sci. 2 (2), 54–64.

Chaudhary, P., Aryal, K.P., 2009. Global warming in Nepal: challenges and policy imperatives. J. For. Livelihood 8 (1), 4–13.

Chaudhuri, A.B., Chaudhury, A., 1994. Mangroves of the Sundarbans: India, vol. 1, first ed. IUCN-The world Conservation Union, Bangkok.

Chauhan, C., 2010. Economies clash over carbon. Hindustan Times, Cancun.

Cheung, A.B.L., 2005. The Politics of Administrative Reforms in Asia: paradigms and legacies, paths and diversities. Governance 18 (2), 257–282.

Chinn, T.J., 1996. New Zealand glacier responses to climate change of the past century. New Zeal. J. Geol. Geophys. 39, 415–428.

Chitale, V.S., Behera, M.D., Roy, P.S., 2014. Future of endemic flora of biodiversity hotspots in India. PLoS One 39, 1–15, doi: 10.1371/journal.pone.0115264.

Chopra, K., Kumar, P., Khan, N.A., 2006. Identifying the economy driven of land use change in mangrove ecosystem: a case study of the Indian Sundarbans. Trade Environment and Rural Poverty, WWF-World Bank Report.

Chowdhury, A., 1987. Mangrove ecosystem of Sundarbans. A Long Term Multidisciplinary Research Approach and Report. Department of Science & Technology, Government of India, Department of Marine Science, University of Calcutta, p. 139.

Church, J.A., White, N.J., 2006. A 20th century acceleration in global sea level rise. Geophys. Res. Lett. 33, 1–4, L01602.

Church, J.A., White, N.J., Konikow, L.F., Domingues, C.M., Cogley, J.G., Rignot, E., Gregory, J.M., van den Broeke, M.R., Monaghan, A.J., Velicogna, I., 2011. Revisiting the earth's sea level and energy budgets from 1961 to 2008. Geophys. Res. Lett. 38, 1–8., L18601.

Ciais, P., Sabine, C., Bala, G., Bopp, L., Brovkin, V., Canadell, J., Chhabra, A., DeFries, R., Galloway, J., Heimann, M., Jones, C., Le Quere, C., Myneni, R.B., Piao, S., Thornton, P., 2013. Carbon and other biogeochemical cycles. Climate Change 2013: The Physical Science Basis. Contribution of Working Group I to the Fifth Assessment Report of the Intergovernmental Panel on Climate Change. Stocker, T.F., Qin, D., Plattner, G.K., Tignor, M., Allen, S.K., Boschung, J., Nauels, A., Xia, Y., Bex, V., Midglay, P.L., (Eds.), Cambridge University Press, Cambridge, UK, pp. 465–570.

Coffin, M.F., Duncan, R.A., Eldholm, O., Fitton, J.G., Frey, F.A., Larsen, H.C., Mahoney, J.J., Saunders, A.D., Schlich, R., Wallace, P.J., 2006. Large igneous provinces and scientific ocean drilling: status quo and a look ahead. Oceanography 19, 150–160.

Cole, D.H., 2015. Advantages of a polycentric approach to climate change policy. Nat. Clim. Chang. 5, 114–118, (doi: 10.1038/nclimate2490).

COMEST, 2010. World Commission on the Ethics of Scientific Knowledge and Technology Report. The Ethical Implication of Global Climate Change. UNESCO, Paris, 75352, pp. 1 -38.

COP 21, 2015. UNFCC, Paris Agreement. Available from: http://unfccc.int/files/essential_background/convention/application/pdf/english_paris_agreement.pdf.

Costello, A., Abbas, M., Allen, A., Ball, S., Bell, S., Bellamy, R., Friel, S., Groce, N., Johnson, A., Kett, M., Lee, M., Levy, C., Maslin, M., McCoy, D., McGuire, B., Montgomery, H., Napier, D., Pagel, C., Patel, J., Antonio, J., de Oliveira, P., Redclift, N., Rees, H., Rogger, D., Scott, J., Stephenson, J., Twigg, J., Wolff, J., Patterson, C., 2009. Managing the health effects of climate change. The Lancet 373 (9676), 1693–1733.

Crutzen, P.J., Ramanathan, V., 2000. The ascent of atmospheric sciences. Science 290 (5490), 299–304.

Cruz, R.V., Harasawa, H., Lal, M., Wu, S., Anokhin, Y., Punsalmaa, B., Honda, Y., Jafari, M., Li, C., Asia, Huu N., 2007. Climate change, 2007, impacts, adaptation and vulnerability. In: Parry, M.L., Canziani, O.F., Palutikof, J.P., et al (Eds.), Contribution of Working Group II to the Fourth Assessment Report of the Intergovernmental Panel on Climate Change. Cambridge University Press, Cambridge, UK, pp. 469–506.

Centre for Science and Environment (CSE), 2012. Report on development deficit to worsen effect of Climate change in Sundarbans. p. 57.

Curray, JR, Emmel, FJ, Moore, DG, Russel, WR, 1982. Structure, tectonics and geological history of the Northwestern Indian Ocean. In: Nairn, AEM, Stehli, FG (Eds.), The Indian Ocean (The Ocean Basins and Margins-6). Plenum, New York, NY, pp. 339–450.

Dagnet, Y., Waskow, D., Elliot, C., Northtrop, E., Thwaits, J., Mogelgaard, K., Krnjic, M., Levin, K., McGray, H., 2016. Staying on Track From Paris: Advancing the Key Elements of the Paris Agreement. Working paper. World Resource Institute, Washington, DC. pp. 1–60.

Datta, D., 2010. Development of a comprehensive environmental vulnerability index for evaluation of the status of Eco Development Committees in the Sundarbans, India. Int. J. South Asian Stud. 3, 258–271.

Davis, C.H., Li, Y., McConnell, J.R., Frey, M.M., Hanna, E., 2005. Snowfall driven growth in East Antarctic ice sheet mitigates recent sea level rise. Science 308, 1898–1901.

Davis, G.T., 2012. Vulnerability Assessment of the Gulf of Khambhat. Dissertation Thesis, National Institute of Oceanography, Goa, India. 78.

De Fraiture, C., Wichelns, D., 2007. Looking ahead to 2050. Scenarios of alternate investment approaches. In: Moblen, D. (Ed.), Water for Food, Water for Life: A Comprehensive Assessment of Water Management in Agriculture. Earth Scan and IWMI, London and Colombo, pp. 279–310.

Dickson, M.H., Fanelli, M., 2004. What is Geothermal Energy? Home page: http://iga.igg.cnr.it/geo/geoenergy.

Dixit, A., 2011. Climate Change in Nepal: Impacts and Adaptive Strategies. World Resources Institute, CA, USA.

Dixon, A., Gonzalez, A., 2017. South Asia's Turn: Policies to Boost Competitiveness and Create the Next Export Powerhouse. International Bank for Reconstruction and Development/The World Bank Group, 1818, H Street, NW, Washington, DC, 20433, pp. 1–16.

Dlugokencky, E.J., Nisbet, E.G., Fisher, R., Lowry, D., 2011. Global atmospheric methane: budget, changes and dangers. Philos. Trans. Royal Soc. A 369, 2058–2072.

Doney, S.C., Fabry, V.J., Feely, R.A., Kleypas, J.A., 2009. Ocean acidification: the other CO2 problem. Annu. Rev. Mar. Sci. 1, 169–192.

Dutta, A., 2016. Estimating Impact of Sea Level Rise in South Asia (Masters Dissertation Thesis), NIO, Goa, 77 p.

Dyurgerov, M.B., Meier, M.F., 1997. Year-to-year fluctuations of global mass balance of small glaciers and their contribution to sea-level changes. Arct. Alp. Res. 29, 392–402.

EAI, 2015. Hydro energy in India. Potential and Future of Hydro Energy in India. Available from: http://www.eai.in/ref/ae/hyd/hyd.html.-2011.

Elliot, J., 2015. Implosion: India's Tryst With Reality. Harper Collins Publishers, India, 507 p.

Encyclopedia, 2017. The world Book of Encyclopedia, www.worldbook.com.

Enzel, Y., Ely, L.L., Mishra, S., Ramesh, R., Amit, R., Lazar, B., Rajaguru, S.N., Baker, V.R., Sandler, A., 1999. High-resolution holocene environmental changes in the Thar Desert, Northwestern India. Science 284 (5411), 125–128.

EM-DAT, 2014. The Emergency Events Database. Universite Catholic de Louvin, CRED, Brussels, Belgium. Available from: www.emdat.be.

Etminan, M., Myhre, G., Highwood, E.J., Shine, K.P., 2016. Radiative forcing of carbon dioxide, methane, and nitrous oxide: a significant revision of the methane radiative forcing. Geophys. Res. Lett. 43, 12614–12623.

European Population Conference (EPC), 2016. Demographic change and policy implications. Mainz, Germany. pp. 1–36.

EU-EDGAR, 2014. Trends in Global CO_2 Emissions: 2014 Report. PBL Netherlands Environmental Assessment Agency, European Union Commission. p. 62.

FAO, 2015. Climate Change and Food Systems: Global Assessment for Food Security and Trade. Food and Agricultural Organization of the United Nations, Rome, pp. 1–356.

FAO, 2012. The State of Food and Agriculture Food and Agricultural Organization of the United Nations, Rome, pp. 1–182.

Farber, D.A., 2007. Adapting to climate change: who should pay. J. Land Use Environ. Law 23, 1–39.

Feely, R.A., Alin, S.R., Newton, J., Sabine, C.L., Warner, M., Devol, A., Krembs, C., Maloy, C., 2010. The combined effects of ocean acidification, mixing, and respiration on pH and carbonate saturation in an urbanized estuary. Estuar. Coast. Shelf Sci. 88, 442–449.

Feely, R.A., Doney, S.C., Cooley, S.R., 2009. Ocean acidification: present conditions and future changes in a high-CO2 world. Oceanography 22, 36–47.

Fidler, D.P., 2010. Asia's Participation in Global Health Diplomacy and Global Health Governance. Articles by Maurer Faculty. Paper 1674. Maurer School of Law, Indiana University, Digital Repository @ Maurer Law. Available from: http://www.repository.law.indiana.edu/facpub/1674.

Fisher, B.S., Nakicenovic, N., Alfsen, K., Corfee Morlot, J., de la Chesnaye, F., Hourcade, J.-Ch., Jiang, K., Kainuma, M., La Rovere, E., Matysek, A., Rana, A., Riahi, K., Richels, R., Rose, S., van Vuuren, D., Warren, R., 2007. Issues related to mitigation in the long-term context. Climate Change 2007: Mitigation. Contribution of Working Group III to the Fourth Assessment Report of the IPCC. Metz, B., Davidson, O.R., Bosch, P.R., Dave, R., Meyer, L.A. (Eds.), Cambridge University Press, Cambridge (Chapter 3).

Fitzharris, 1996. The cryosphere: changes and their impacts. In: Houghton, J.T., Filho, L.G.M., Callander, B.A., Harris, N., Kattenberg, A., Maskell, K. (Eds.), In Climate Change 1995: The Science of Climate Change. Cambridge University Press, Cambridge, UK, 241-265.

Fourier, J.B.J., 1824. Remarques générales sur les températures du globe terrestre et des espaces planétaires. Annales de Chimie et de Physique 27, 136–167.

Frank, J., 2015. Explaining how the water vapour greenhouse effect works. Skeptical Science, 1–3, SkepticalScience.com.

Galbraith, J., 2011. To what extent is business responding to climate change? evidence from a global wine producer. J. Bus. Ethics 104 (3), 421–432.

Ganguly, D., Mukhopadhyay, A., Pandey, R.K., Mitra, D., 2006. Geomorphological study of Sundarbans deltaic estuary. Indian Soc. Remote Sens. 34, 431–435.

Gardiner, S., 2016. Why climate change is an ethical problem. The Washington Post.

Ghosh, T., Bhandari, G., Hazra, S., 2003. Application of a bio-engineering technique to protect Ghoramara Island (Bay of Bengal) from severe erosion. J. Coast. Conserv. 9, 171–178.

Gibbert, P., van Kilfschoten, T., 2004. The pleistocene and holocene epochs. In: Gradstein, F.M., Ogg, J.G., Gilbert, S.A. (Eds.), Geologic Time Scale 2004. Cambridge University Press, Cambridge, Chapter 22.

Glaser, B., 2007. Prehistorically modified soils of central Amazonia: a model for sustainable agriculture in the twenty-first century. Philos. Trans. 362, 187–196.

Gopal, B., Chauhan, M., 2006. Biodiversity and its conservation in the Sundarbans mangrove ecosystem. Aquatic Sci. 68, 338–354.

Gordon, D., 2017. Understanding Climate Engineering. Carnegie Endowment for International peace. The Global Think Tank, CarnegieEndowment.org.

Gorin, P., Clot, N., 2016. Model 5: Mitigation of Climate Change in Afghanistan. Helvetas Swiss Intercooperation, Afghanistan. p. 6.

Goswami, D.Y., Besarati, S.M., 2013. World Energy Resources: Solar Energy in World Energy Council 2013. World Energy Council, UK, pp. 325–468 (Chapter 8).

Green, D., Alexander, L., Mclnnes, K., Church, J., Nicholls, N., White, N., 2009. An assessment of climate change impacts and adaptation for the Torres Strait Islands, Australia. Clim. Change 102 (3-4), 405–433.

Greenpeace European Renewable Energy Council, 2007. Energy Revolution: A Blueprint for Solving Global Warming. Available from: http://www.energyblueprint.info.

Geological Survey of India (GSI) Report, 2004. Plate tectonics, earthquake and tsunami in Sundarbans Delta Complex. Available from: http://www.portal.gsi.gov.in.

Gumartini, T., 2009. Biomass energy in the Asia-Pacific region: current status, trends and future setting. Food and Agricultural Organization of the United Nations Regional Office for Asia and the Pacific. Asia-Pacific Forestry Sector Outlook Study II, Working Paper Series, Working Paper No. APFSOS II/WP/2009/26, Bangkok, pp. 1–46.

GWEC – Global Wind Energy Council, 2012. Wind Energy Future in Asia: A Compendium of Wind Energy Resource Maps, Project Data and Analysis for 17 Countries in Asia and the Pacific, pp. 1–104, Full Report August 2012.

Haeberli, W., Beniston, M., 1998. Climate change and its impacts on glaciers and permafrost in the Alps. Ambio 27, 258–265.

Hall, D., 2015. Why Public–Private Partnerships don't work. The Many Advantages of the Public Alternative. PSIRU (Public Service International Research Unit), University of Greenwich, UK, pp. 1–56.

Hallegatte, S., Bangalore, M., Bonzanigo, L., Fay, M., Kane, T., Narloch, U., Rozenberg, J., Treguer, D., Vogt-Schilb, A., 2016. Shock waves: managing the impacts of climate change on poverty. Climate Change and Development Series. Washington, DC, World Bank.

Haller, S.F., Gerrie, J., 2011. The role of science in public policy: higher reason, or reason for hire? J. Agric. Environ. Ethics 20, 139–165.

Hall-Spencer, J.M., Rodolfo-Metalpa, R., Martin, S., Ransome, E., Fine, M., Turner, S.M., Rowley, S.J., Tedesco, D., Buia, M.C., 2008. Volcanic carbon dioxide vents show ecosystem effects of ocean acidification. Nature 454, 96–99.

Hansen, M.C., Potapov, P.V., Moore, R., Hancher, M., Turubanova, S.A., Tyukavina, A., Thau, D., Stehman, S.V., Goetz, S.J., Loveland, T.R., Kommareddy, A., Egorov, A., Chini, L., Justice, C.O., Townshend, J.R.G., 2013. High-resolution global maps of 21st-century forest cover change. Science 342, 850–853.

Hansen, J., Sato, M., Kharecha, P., Beerling, D., Berner, R., Masson-Delmotte, V., Pagani, M., Raymo, M., Royer, D.L., Zachos, J.C., 2008. Target atmospheric CO_2: where should humanity aim? Open Atmos. Sci. J. 2, 217–231.

Haque, M.S., 2001. Recent transition in governance in South Asia: context, dimensions and implications. Int. J. Public Admin. 24 (12), 1405–1436.

Hardy, J.T., 2003. Climate Change: Causes Effects and Solutions. John Wiley & Sons Limited, pp. 247.

Harrabin, R., 2007. China building more power plants. Available from: http://news.bbc.co.uk/1/hi/world/asia-pacific/6769743.stm.

Harris, C., Mann, R., 2014. Global temperature from 2500BC to 2040AD. Long-Range Weather. Available from: www.LongRangeWeather.com.

Hascic, I., Watson, F., Johnstone, N., Kaminker, C., 2012. Recent trends in innovation in climate change mitigation technologies. Energy and Climate Policy, pp. 17–53.

Hastenrath, S., Greischar, L., 1997. Glacier recession on Kilimanjaro, East Africa 1912–1989. J. Glaciol. 43, 455–459.

Hayhoe, K., 2015. Science and Faith Can Solve Climate Change Together. In: Do or Die: The Global Climate Summit in Paris, Scientific American (25 November, 2015).

HDR, 2006. Human Development Report 2006. Beyond Scarcity: Power, Poverty and the Global Water Crisis. United Nations Development Programme, New York, NY.

HDR (Human Resource Development), 2016. United Nations Development Program. Available from: www.hdr.undp.org.

Hegerl, G.C., Zwiers, F.W., Braconnot, P., Gillett, N.P., Luo, Y., Marengo Orsini, J.A., Nicholls, N., Penner, J.E., Stott, P.A., 2007. Understanding and Attributing Climate Change. In: Climate Change 2007: The Physical Science Basis. Contribution of Working Group I to the Fourth Assessment Report of the Intergovernmental Panel on Climate Change. Solomon, S., Qin, D.,

Manning, M., Chen, Z., Marquis, M., Averyt, K.B., Tignor, M., Miller, H.L. (Eds.). Cambridge University Press, Cambridge, UK; New York, NY, USA.

Hinds, W.C., 1999. Aerosols technology: properties, behavior and measurement of air-borne particles, second ed. A Wiley Interscience Publication, New York, NY.

Hinkel, J, Lincke, D., Vafeidis, A.T., Perrette, M., Nicholls, R.J., Tol, R.S.J., Marzeion, B., Fettweis, X., Ionescu, C., Levermann, A., 2014. Coastal flood damage and adaptation costs under 21st century sea-level rise. Proceedings of National Academy of Sciences, vol. 111. pp. 3292–3297.

Hirway, I., Goswami, S., 2007. Valuation of Coastal Resources: The Case of Mangroves in Gujarat. Academic Foundation, New Delhi, p. 170.

Hofstede, G., 1994. The business of international business is culture. Int. Bus. Rev. 3 (1), 1–14.

Houghton, J.T., Ding, Y., Griggs, D.J., Noguer, M., van der Linden, P.J., Dai, X., et al., 2001. Climate Change 2001: The Scientific Basis. IPCC, Cambridge University Press, Cambridge, p. 39.

Howes, S., Wyroll, P., 2012. Climate Change Mitigation and Green Growth in Developing Asia. ADBI Working Paper 369. Asian Development Bank Institute, Tokyo.

Hussain, K., 2006. A vision for South Asia. Indian J. Politics Int. Relat., 1–17.

ICMAM, 2002. Integrated Coastal and Marine Area Management Report. Ministry of Earth Sciences, New Delhi.

Idso, C., Singer, S.F., 2010. Climate Change Reconsidered: 2009 Report of the Non-Governmental International Panel on Climate Change, NIPCC. The Heartland Institute, USA, pp. 754.

Idso, C.D., Idso, S.B., Carter, R.M., Singer, S.F. (Eds.), 2014. Climate Change Reconsidered II: Biological Impacts. The Heartland Institute, Chicago, IL.

IEA, 2016. International Energy Agency: Key World Energy Statistics. IEA, Paris. 75739, p. 80.

IEA-REN, 2016. Renewables 2016. Global Status Report. REN (Renewable Energy Policy Network for 21st Century)-21. REN21 Secretariat,Paris, France, pp. 1–272.

IFRC – International Federation of Red Cross and Red Crescent Societies, 2009. Climate Change Adaptation, Strategies for Local Impact: Key Messages for UNFCCC Negotiators. International Federation of Red Cross and Red Crescent Societies, pp. 1–11.

Iftikhar, M.N., Najeeb, F., Mohazzam, S., Khan, S.A., 2015. Sustainable Energy for All in South Asia: potential, challenges and solutions. Sustainable Development Policy Institute (SDPI), Islamabad, Pakistan. Working Paper No. 151. pp. 1–33.

International Institute for Environment and Development (IIED), 2011. Mainstreaming Environment and Climate Change. Agriculture Briefing, April. pp. 1–8. Available from: http://pubs.iied.org/pdfs/G03098.pdf.

IMD-Customized Rainfall Information System India Meteorological Department (CRIS-IMD), 2012. Ministry of Earth Sciences, New Delhi, India.

International Monetary Fund (IMF), 2015. Islamic Republic of Afghanistan, IMF Country Report No. 15/324, p. 112.

INDC-India, 2015. India's Intended Nationally Determined Contribution: Working Towards Climate Justice Submitted to UNFCCC, p. 38.

IPCC, 1996. Climate Change 1995: The Science of Climate Change. In: Houghton, J.T., Meira Filho, L.G., Callander, B.A., Harris, N., Kattenberg, A., Maskell, K. (Eds.), Contribution of WGI to the Second Assessment Report of the Intergovernmental Panel on Climate Change. Published for the Intergovernmental Panel on Climate Change, Cambridge University Press, The Pitt Building, Trumpington Street, Cambridge, UK, pp. 1–588.

IPCC, 1990. Climate Change: The IPCC scientific assessment. In: Houghton, J.T., Jenkins, G.J., Ephraums, J.J. (Eds.), Contribution of WGI to the First Assessment Report of the Intergovernmental Panel on Climate Change. Published for the Intergovernmental Panel on Climate

Change, Cambridge University Press, The Pitt Building, Trumpington Street, Cambridge, UK. pp. 1–414.

IPCC, 2001. Climate change 2001. In: Houghton, J.T., Ding, Y., Griggs, D.J., Noguer, M., van der Linden, P.J., Dai, X., Maskel, K., Johnson, C.A. (Eds.), The Scientific Basis. Contribution of the Working Group I to the Third Assessment Report of the Intergovernmental Panel on Climate Change. Cambridge University Press, Cambridge, UK, New York, NY, USA, pp. 881.

IPCC, 2007. 4. Adaptation and mitigation options. Summary for Policymakers. Climate Change 2007: Synthesis Report. Contribution of Working Groups I, II and III to the Fourth Assessment Report of the Intergovernmental Panel on Climate Change (Core Writing Team, Pachauri, R.K., Reisinger, A. (Eds.)). IPCC, Geneva, Switzerland.

Intergovernmental Panel on Climate Change (IPCC), 2013. Climate Change 2013: The Physical Science Basis. In: Stocker, T.F., Quinn, D., Plattner, G.K., Tignor, M., Allen, S.K., Boschung, J., Nauels, A., Xia, Y., Bex, V., Midgley, P.M. (Eds.), Contribution of Working Group I to the Fifth Assessment Report of the Intergovernmental Panel on Climate Change. Cambridge University Press, Cambridge, UK, pp. 1–33.

IPCC, 2014. Climate Change 2014: Impacts, Adaptation, and Vulnerability. Part A: Global and Sectoral Aspects. In: Field, C.B., Barros, V.R., Dokken, D.J., Mach, K.J., Mastrandrea, M.D., Bilir, T.E., Chatterjee, M., Ebi, K.L., Estrada,Y.O., Genova, R.C., Girma, B., Kissel, E.S., Levy, A.N., MacCracken, S., Mastrandrea, P.R., White, L.L. (Eds.), Contribution of Working Group II to the Fifth Assessment Report of the Intergovernmental Panel on Climate Change. Cambridge University Press, Cambridge, UK, New York, NY, USA, pp. 1132.

International Union for Conservation of Nature (IUCN), 2012. Annual Report, Nature and Towards Nature-Based Solutions. IUCN, World headquarters, Switzerland, p. 36.

Ives, J.D., Shreshta, R.B., Mool, P.K., 2010. Formation of Glacial Lakes in the Hindu Kush-Himalayas and GLOF Risk Assessment. International Center for Integrated Mountain Development (ICIMOD), Kathmandu, Nepal, ICIMOD, UNSDR, GFDRR.

Jacoby, H.D., Janetos, A.C., Birdsey, R., Buizer, J., Calvin, K., de la Chesnaye, F., Schimel, D., Sue Wing, I., Detchon, R., Edmonds, J., Russell, L., West, J., 2014. Ch. 27: Mitigation. Climate Change Impacts in the United States. In: Melillo, J.M., Richmond, T., Yohe, G.W. (Eds.), The Third National Climate Assessment. US Global Change Research Program, pp. 648–669. doi:10.7930/J0C8276J.

Jamil, I., Askvik, S., Dhakal, T.K., 2013. Understanding governance in South Asia. In: Jamil, I., Askvik, S., Dhakal, T.K. (Eds.), In Search of Better Governance in South Asia and Beyond, Public Administration, Governance and Globalization. Springer Science, Business Media, New York, NY, doi: 10.1007/978-1-4614-7372-5-2.

Jansen, E., Overpeck, J., Briffa, K.R., Duplessy, J.C., Joos, F., Masson-Delmotte, V., Olago, D., Otto-Bliesner, B., Peltier, W.R., Rahmstorf, S., Ramesh, R., Raynaud, D., Rind, D., Solomina, O., Villalba, R., Zhang, D., 2007 Climate Change: The Physical Science Basis. Working Contribution of Working Group I to the Fourth Assessment Report of the Intergovernmental Panel on Climate Change. pp. 433–498.

Janssens-Maenhout, G., Crippa, M., Guizzardi, D., Muntean, M., Schaaf, E., Olivier, J.G.J., Peters, J.A.H.W., Schure, K.M., 2017. Fossil CO_2 and GHG emissions of all world countries, EUR 28766 EN. Publications Office of the European Union, Luxembourg, ISBN: 978-92-79-73207-2, doi:10.2760/709792, JRC107877.

Jaramillo, C., Ochoa, D., Contreras, L., Pagani, M., Carvajal-Ortiz, H., Pratt, L.M., Krishnan, S., Cardona, A., Romero, M., Quiroz, L., Rodriguez, G., Rueda, J.M., de la Parra, F., Morón, S., Green, W., Bayona, G., Montes, C., Quintero, O., Ramirez, R., Mora, G., Schouten, S., Bermudez, H., Navarrete, R., Parra, F., Alvarán, M., Osorno, J., Crowley, J.L., Valencia, V., Vervoort, J., 2010. Effects of rapid global warming at the Paleocene-Eocene Boundary on neotropical vegetation. Science 330, 957–961.

Jayappa, K.S., Mitra, D., Mishra, A.K., 2006. Coastal geomorphological and land-use and land-cover study of Sagar Island, Bay of Bengal (India) using remotely sensed data. Int. J. Remote Sens 27, 3671–3682.

Jilani, T., Gomi, K., Matsuoka, Y., 2012. Low-Carbon Society Development Towards 2025 in Bangladesh. Asia Pacific Integrated Model, Kyoto University, Japan, pp. 10–48.

Karl, T.R., Melillo, J.T., Peterson, T.C. (Eds.), 2009. Global Climate Change Impacts in the United States. Cambridge University Press, p. 189.

Kaur, S., Kaur, H., 2017. Climate Change Begs for Policy Initiatives in South Asia. East Asia Forum. Available from: http://www.eastasiaforum.org/2017/08/26/climate-change-begs-for-policy-initiatives-in-south-asia/, pp. 1–3.

Keith, D.W., 2010. Photophoretic levitation of engineered aerosols for geoengineering. Proceedings of National Academy of Sciences,vol. 107, no. 38. pp. 16428–16431.

Keith, D.W., 2000. Geo-engineering the climate: history and prospect. Annu. Rev. Energy Environ. 25, 245–284.

Kelkar, U., Bhadwal, S., 2007. South Asian Regional Study on Climate Change Impacts and Adaptation: Implications for Human Development. Human Development Report Office, Occasional Paper. United Nations Development Programme.

Kennedy, J.J., Thorne, P.W., Peterson, T.C., Reudy, R.A., Stott, P.A., Parker, D.E., Good, S.A., Titchner, H.A., Willett, K.M., 2010. How do we know the world has warmed? [in "State of the Climate in 2009"]. Bull. Am. Meteorol. Soc. 91, S26–27.

Khalil, M.A.K., 1999. Non-CO2 Greenhouse gases in the atmosphere. Annu. Rev. Energy Environ. 24, 645–661.

Khan, S., 2012. Effects of climate change on Thatta and Badin. Available from: envirocivil.com.

Khan, I., Chowdhury, H., Rasjidin, R., Alam, F., Islam, T., Islam, S., 2012. Review of Wind energy utilization in South Asia. Procedia Eng 49, 213–220.

Khan, R., Shahjahan, M., 2014. Low Carbon South Asia: Bangladesh. Christian Aid, London, p. 23.

Kiehl, J.T., Trenberth, K.E., 1997. Earth's annual global mean energy budget. Bull. Am. Meteorol. Soc. 78, 197–208.

Kim, J., Fraser, P.J., Li, S., Mühle, J., Ganesan, A.L., Krummel, P.B., Steele, L.P., Park, S., Kim, S.K., Park, M.K., Arnold, T., Harth, C.M., Salameh, P.K., Prinn, R.G., Weiss, R.F., Kim, K.R., 2014. Quantifying aluminium and semiconductor industry perfluorocarbon emissions from atmospheric measurements. Geophys. Res. Lett. 41, 4787–4794.

Klimont, Z., Smith, S.J., Cofala, J., 2013. The last decade of global anthropogenic sulphur dioxide: 2000–2011 emissions. Environ. Res. Lett. 8, 1–7.

Kopp, R.E., Kirschvink, J.L., Hilburn, I.A., Hilburn, F., Nash, C.Z., 2005. The Paleoproterozoic snow ball Earth: a climate disaster triggered by the evolution of oxygenic photosynthesis. Proc. Natl. Acad. Sci. 102, 11131–11136.

Kotal, S.D., Kundu, P.K., Roy Bhowmik, S.K., 2009. An analysis of sea surface temperature and maximum potential intensity of tropical cyclones over the Bay of Bengal between 1981 and 2000. Meteorol. Appl. 16, 169–180.

Kutterolf, S., Jegen, M., Mitrovica, J.X., Kwasnitschka, T., Freundt, A., Huybers, P.J., 2013. A detection of Milankovitch frequencies in global volcanic activity. Geology 41, 227–230.

Laborde, D., 2011. Climate change and agriculture in south Asia: Looking for an optimal trade policy. Selected Paper Prepared for Presentation at the Agricultural and Applied Economics Association's 2011 AAEA and NAREA Joint Annual Meeting. Pittsburg, Pennsylvania, July 24–26, 2011, pp. 1–58.

Lacis, A.A., Schmidt, G.A., Rind, D., Ruedy, R.A., 2010. Atmospheric CO2: principal control knob governing earth's temperature. Science 330, 356–359.

Lal, R., 2004. Soil carbon sequestration impacts on global climate change and food security. Science 304, 1623–1627.

Lal, R., 2007. Soil degradation and environment quality in South Asia. Int. J. Ecol. Environ. Sci. 33 (2–3), 91–103.

Lal, M., 2003. Global climate change: India's monsoon and its variability. J. Environ. Stud. Policy 6, 1–34.

Lali, A., 2016. Biofuels for India: what, when and how? Curr. Sci. 110 (4), 552–555.

Lamarque, J.F., Hess, P., Emmons, L., Buja, L., Washington, W., Granier, C., 2005. Tropospheric ozone evolution between 1890 and 1990. J. Geophys. Res. 110, 1–15, D08304, doi:10.1029/2004JD005537.

Langmuir, C., Broecker, W., 2012. How to Build a Habitable Planet: The Story of Earth From the Big Bang to Humankind. Princeton University Press, p. 718.

Laurance, W.F., Williamson, G.B., 2001. Positive feedbacks among forest fragmentation, drought, and climate change in the Amazon. Conserv. Biol. 15, 1529–1535.

Letmathe, P.B., Biswas, A.K., 2015. Very few drops to drink: water pollution is a bigger environmental threat than climate change. The Times of India.

Le Quere, C., Raupach, M.R., Canadel, J.G., Marland, G., Bopp, L., Ciais, P., Conway, T.J., Doney, S.C., Feely, R.A., Foster, P., Friedlingstein, P., Gurney, K., Houghton, R.A., House, J.I., Huntingford, C., Levy, P.E., Lomas, M.R., Majkut, J., Metzel, N., Ometto, J.P., Peters, G.P., Prentice, I.C., Randerson, J.T., Running, S.W., Sarmiento, J.L., Schuster, U., Sitch, S., Takahasi, T., Viovy, N., van der Werf, G.R., Woodward, F.I., 2009. Trends in the sources and sinks of carbon dioxide. Nat. Geosci. 2, 831–836.

Levine, M., Ürge-Vorsatz, D., Blok, K., Geng, L., Harvey, D., Lang, S., Levermore, G., Mongameli Mehlwana, A., Mirasgedis, S., Novikova, A., Rilling, J., Yoshino, H., 2007. Chapter 6: Residential and commercial buildings. In: Metz, B., Davidson, O.R., Bosch, P.R., Dave, R., Meyer, L.A., (Eds.), Climate Change 2007: Working Group III: Mitigation of Climate Change, l"Sec 6.4.2 Thermal envelope ,Contribution of Working Group III to the Fourth Assessment Report of the Intergovernmental Panel on Climate Change, Cambridge University Press, Cambridge, United Kingdom and New York, NY, USA.

Liberty Institute New Delhi, 2002. World summit for sustainable Development, August-September, 2002, Julian L Simon Centre, New Delhi, India.

Lockwood, J.G., 1979. Causes of Climate. Halsted Press, John Wiley & Sons, New York, NY.

Lohani, S.P., Baral, B., 2012. Conceptual framework of low carbon strategy for Nepal. Low Carbon Econ. 2, 230–238.

Lovejoy, S., Schertzer, D., 2013. The weather and climate: emergent laws and multi-fractal cascades. Cambridge University Press, p. 496.

Lu Q.B. 2013. Cosmic-ray-driven reaction and green house gas effect of halogenated molecules: culprits of atmospheric ozone depletion and global climate change. Int. J. Mod. Phys. B, 27, 1350073 (2013) [38 pages] https://doi.org/10.1142/S0217979213500732.

Lu, Z., Zhang, Q., Streets, D.G., 2011. Sulfur dioxide and primary carbonaceous aerosol emissions in China and India, 1996-2010. Atmos. Chem. Phys. 11, 9839–9864.

Lunt, M.F., Rigby, M., Ganesan, A.L., Manning, A.J., Prinn, R.G., O'Doherty, S., Mühle, J., Harth, C.M., Salameh, P.K., Arnold, T., Weiss, R.F., Saito, T., Yokouchi, Y., Krummel, P.B., Steele, L., Fraser, P.J., Li, S., Park, S., Reimann, S., Vollmer, M.K., Lunder, C., Hermansen, O., Schmidbauer, N., Maione, M., Arduini, J., Young, D., Simmonds, P.G., 2015. Reconciling reported and unreported HFC emissions with atmospheric observations. Proc. Natl. Acad. Sci. 12 (19), 5927–5931.

Lwasa, S., 2015. A systematic review of research on climate change adaptation policy and practice in Africa and South Asia deltas. Reg. Environ. Change 15, 815–824.

Mall, R.K., Kumar, S., 2014. Integration of Disaster Reduction and Climate Change Adaptation in SAARC Countries (Implementation of Thimpu Statement on Climate Change, UNISDR/ SAARC-DMC/ BHU), Varanasi, 260 p.

Manzoor, R., Ramay, S.A., 2013. Green Growth and Technological Innovation: A Case for South Asian countries, Working Paper # 136. Environment and Climate Change Unit, Sustainable Development Policy Institute (SDPI), Islamabad, pp. 1–17.

Marcu, A., Stoefs, W., Belis, D., Katja, T., 2015. Country Case Study—Maldives Climate for Sustainable Growth. Center for European Policy Studies (CEPS), Brussels, pp. 51.

Martin, G., 2010. Human values and ethics in workplace. Glenn Martin, 317 pp. ISBN: 978-0-9804045-0-0.

Mazar, M.S., Goraya, N.S., 2015. Issues of good governance in South Asia. South Asian Studies 30 (2), 125–160.

McGranahan, G., Balk, D., Anderson, B., 2007. The rising tide: assessing the risks of climate change and human settlements in low elevation coastal zones. Environ. Urban. 19 (1), 17–37.

McKinsey Quarterly, 2010. The Business Opportunity in Water Conservation, Number 1. Available from: http://aquadoc.typepad.com/files/mckinsey-the-business-of-water-dec-09.pdf, 2010.

MEEGM – Ministry of Environment and Energy Government of Maldives, 2015. Maldives Intended Nationally Determined Contribution, p. 12.

Meehl, G.A., Aixue Hu, A., Arblaster, J.M., Fasullo, J., Trenberth, K.E., 2013. Externally forced and internally generated decadal climate variability associated with the inter-decadal Pacific Oscillation. J. Clim. 26, 7298–7310.

Meisen, P., Azizy, P., 2008. Rural Electrification in Afghanistan, How Do We Electrify the Villages of Afghanistan? Global Energy Network Institute (GENI), CA, USA, pp. 1–26.

MENR (Ministry of Environment and Natural Resources), 2000. Initial National Communication Under the United Nations Framework Convention on Climate Change: Sri Lanka. MENR, Government of Sri Lanka, Colombo.

Menzel, C., 2017. World Energy Scenarios: The grand transition. APERC Annual Conference, Tokyo Japan, 1-13

Merrey, D.J., Prakash, A., Swatuk, L., Jacobs, I., Narain, V., 2017. Chapter 12: Water governance futures in South Asia and Southern Africa: Déjà Vu all over again? In: Karar, E., (Ed.), Freshwater Governance for the 21st Century, Global Issues in Water Policy 6, pp. 229–250. doi: 10.1007/978-3-319-43350-9_1.

Mesagen, E.P., 2015. Economic Growth and Carbon Emission in Nigeria, pp. 1–16. Available from: https://www.researchgate.net/publication/307575889.

Michels, K.H., Kudrass, H.R., Hübscher, C., Suckow, A., Wiedicke, M., 1998. The submarine delta of the Ganges-Brahmaputra: cyclone-dominated sedimentation patterns. Mar. Geol. 149, 133–154.

Milankovitch, M.M., 1941. Kanon der Erdbestrahlung. Beograd. Koninglich Serbische akademie, (English translation: Canon of Insolation and the Ice Age Problem, by Israel Program for Scientific Translation and published for the US Department of Commerce and the National Science Foundation), pp. 484.

Miller, B., 2015. Five things you need to know about COP21. CNN Meteorologist, CNN News.

Milliman, J.D., Meade, R.H., 1983. World-wide delivery of river sediments to the oceans. J. Geol. 91, 1–22.

Ming, T., de Richter, R., Liu, W., Caillol, S., 2014. Fighting global warming by climate engineering: is the earth radiation management and the solar radiation management any option for fighting climate change? Renew. Sustain. Energy Rev. 31, 792–834.

Ministry of Environment and Construction, 2005. State of the Environment: Maldives 2005. Ministry of Environment and Construction, Maldives.

Mirza, M., 2007. Climate Change: The Physical Science Basis. Adaptation and Impacts Research Division (AIRD) Environment Canada. C/o-Department of Physical and Environmental Sciences, University of Toronto at Scarborough, 1265 Military Trail, Toronto, Ontario M1C 1A4, Canada.

Mirza, M.M.Q., 1998. Diversion of Ganges water at Farakka and its effect on the salinity of Bangladesh. J. Environ. Manage. 22, 711–722.

Mirza, M.M.Q., 2002. Global warming and changes in the probability of occurrence of floods in Bangladesh and implications. Glob. Environ. Change 12, 127–138.

Mirza, M.M.Q., Dixit, A., 1997. Climate Change in GBM Basins. Water Nepal 5, Kathmandu, Nepal.

Moench, M., Dixit, A., 2004. Adaptive Capacity and Lively-Hood Resilience: Adaptive Strategies for Responding to Floods and Droughts in South Asia. Institute for Social and Environmental Transition (ISET), Boulder, Nepal, Kathmandu.

Mora, C., Wei, C.L., Rollo, U., Amaro, T., Baco, A.R., Billet, D., Bopp, L., Chen, Q., Collier, M., Danovaro, R., Gooday, A.J., Grupe, B.M., Halloran, P.R., Ingels, J., Jones, D.O.B., Levin, L.A., Nakano, H., Norling, K., Ramirez-Llorda, E., Rex, M., Ruhl, H.A., Smith, C.R., Sweetman, A.K., Thurber, A.R., Tjiputra, J.F., Usseglio, P., Watling, L., Wu, T., Yasuhara, M., 2013. Biotic and human vulnerability to projected changes in ocean biogeochemistry over the 21st century. PLoS Biol. 11 (10), e1001682.

Morello, L., 2010. Ocean acidification threatens global fisheries. Climatewire Sci. Am. 303 (6).

Morner, N.A., Etiope, G., 2002. Carbon degassing from the lithosphere. Glob. Planet. Change 33 (1-2), 185–203.

Mosley-Thompson, E., 1997. Glaciological Evidence of Recent Environmental Changes. Annual Meeting of the Association of American Geography. Fort Worth, TX.

Muis, S., Verlaan, M., Winsemius, H.C., Aerts, J.C.J.H., Ward, P.J., 2016. A global reanalysis of storm surges and extreme sea levels. Nat. Commun. 7, 11969.

Mukhopadhyay, R., 2008. Interlinking of Rivers: Are we on the Right Side? Goa Today. XLIII 3, 36–38.

Mukhopadhyay, R., 2018. Global Warming: Formulating a Science Based Public Policy and Mitigation Strategy for South Asia. Department of Public Administration, School of Social Sciences, Indira Gandhi National Open University, Delhi, 150 p.

Mukhopadhyay, R., George, P., Ranade, G., 1997. Spreading rate dependent seafloor deformation in response to India-Eurasia collision: results of a hydrosweep survey in the Central Indian Ocean Basin. Mar. Geol. 140, 219–229.

Mukhopadhyay, R., Karisiddaiah, S.M., Ghosh, A.K., 2012. Geodynamics of the Amirante Ridge and Trench complex, western Indian Ocean. Int. Geol. Rev. 54, 81–92.

Mukhopadhyay, S.K., Biswas, H., De, T.K., Jana, T.K., 2006. Fluxes of nutrients from the tropical River Hooghly at the land-ocean boundary of Sundarbans, NE Coast of Bay of Bengal. Indian J. Mar. Sci. 62, 9–21.

Munawar, S., 2016. Bhutan Improves Economic Development as a Net Carbon Sink. Climate Institute, Washington, DC, p. 16.

Nachmany, M., Fankhauser, S., Davidova, J., Kingsmill, N., Landesman, T., Roppongi, H., Schleifer, P., Setzer, J.,Sharman, A., Singleton, C.S., Sundaresan, J., Townsend, T., 2015. Climate change legislation in Bangladesh; An excerpt from The 2015 Global Climate Study: a review of climate change legislation in 99 countries. Grantham Research Institute on Climate Change and the Environment, p. 11.

Namgyel, T., Sonam Dagay, C.C.U., Tashi, T., Khandu, S.L., Tshering, K., 2011. Second National Communication from Bhutan to the UNFCCC. National Environment Commission, Royal Government of Bhutan, Thimpu, Bhutan.

Nandy, D., Munoz-Jaramillo, A., Martens, P.C.H., 2011. The unusual minimum of sunspot cycle 23 caused by meridional plasma flow variations. Nature 471, 80–82.

NAPCC, 2008. National Action Plan on Climate Change. Available from: http://pmindia.nic.in/Pg01-52.pdf.

NASA, 2014. Warmest Year in Modern Record. Global Climate Change, National Aeronautics and Space Administration, USA.

NAS, 2015a. Committee on Geoengineering Climate (CGCa). Climate Intervention: Carbon Dioxide Removal and Reliable Sequestration. National Academy of Sciences, The National Academies Press, Washington, DC, USA. p. 154.

NAS, 2015b. Climate Intervention: Reflecting Sunlight to Cool Earth. National Academy of Sciences, National Academies Press, Washington, DC, USA, p. 234.

National Greenhouse Gas Inventory 2008. Pakistan.

NCA, 2014. 2014-National Climate Assessment: Overview. Climate Change Impacts in the United States. US Global Change Research Program. pp. 1–11.

NCCP, 2013. Framework for Implementation of Climate Change Policy (2014-2030). Climate Change Division, Government of Pakistan, Islamabad, pp. 93.

NCDC-NOAA, 2016. NOAA National Centers for Environmental Information, State of the Climate: Global Analysis.

NCVST – Nepal Climate Vulnerability Study Team, 2009. Vulnerability Through the Eyes of Vulnerable: Climate Change Induced Uncertainties and Nepal's Development Predicaments. Institute for Social and Environmental Transition (ISET); Institute for Social and Environmental Transition (ISET-N), Boulder, CO; Kathmandu, Nepal. pp. 1–114.

NEPA, 2013. Greenhouse Gas Emission Future for Afghanistan, Available from: http://www.unep.org/disastersandconflicts/CountryOperations/Afghanistan/News/AfghanistanLEDS/tabid/1060395/Default.aspx.

NEEM, 2013. North Greenland Eemian Ice Drilling. Eemian interglacial reconstructed from Greenland folded ice core. Nature 493, 489–494.

Nepstad, D.C., Boyd, W., Stickler, C.M., Bezerra, T., Azevedo, A.A., 2017. Responding to climate change and the global land crisis: REDD+, market transformation and low-emissions rural development. Phil. Trans. R. Soc. B 368 (20120167), pp. 1–13.

Nicholls, R.J., Cazenave, A., 2010. Sea-level rise and its impact on coastal zones. Science 238, 1517–1520.

Nikolov, N., Zeller, K., 2011. Unified Theory of Climate: Expanding the Concept of Atmospheric Greenhouse Effect Using Thermodynamic Principles: Implications for Predicting Future Climate Change. A Poster at Open Science Conference of the World Climate Research Program, 24 October 2011, Denver, CO, USA.

NIPCC, 2010. Climate Change Reconsidered: The Report of the Non-Governmental International Panel on Climate Change. Heartland Institute, pp. 756.

Northcott, M.S., 2007. A Moral Climate: The Ethics of Global Warming. Darton, Longman & Todd Ltd, London, UK, pp. 1–325.

NOAA, 2016. Global Climate Report. National Centre for Environmental information, National Oceanic & Atmospheric Administration, Asheville, USA.

NOAA-SIO, 2013. Selectivity: Theory, Estimation and Application in Fishery Stock Assessment Model. National Oceanic & Atmospheric Administration – Scripps Institution of Oceanography, Workshop Series Part I, p. 50.

Oppenheimer, M., Campos, M., Warren, R., Birkmann, J., Luber, G., O'Neill, B., Takahashi, K., 2014. Section 19.7.1: Relationship between adaptation efforts, mitigation efforts, and residual impacts. Chapter 19: Emergent risks and key vulnerabilities. In: Field, C.B., Barros, V.R.,

Dokken, D.J., Mach, K.J., Mastrandrea, M.D., Bilir, T.E., Chatterjee, M., Ebi, K.L., Estrada, Y.O., Genova, R.C., Girma, B., Kissel, E.S., Levy, A.N., MacCracken, S., Mastrandrea, P.R., White, L.L. (Eds.), Climate Change 2014: Impacts, Adaptation, and Vulnerability. Part A: Global and Sectoral Aspects. Contribution of Working Group II to the Fifth Assessment Report of the Intergovernmental Panel on Climate Change. Cambridge University Press, Cambridge, UK, New York, NY, USA. pp. 1080–1083.

Orr, J.C., Fabry, V.J., Aumont, O., Bopp, L., Doney, S.C., Feely, R.A., Gnanadesikan, A., Gruber, N., Ishida, A., Joos, F., Key, R.M., Lindsay, K., Maier-Reimer, E., Matear, R., Monfray, P., Mouchet, A., Najjar, R.G., Plattner, G.K., Rodgers, K.B., Sabine, C.L., Sarmiento, J.L., Schlitzer, R., Slater, R.D., Totterdell, I.J., Weirig, M.F., Yamanaka, Y., Yool, A., 2005. Anthropogenic ocean acidification over the twenty-first century and its impact on calcifying organisms. Nature 437, 681–686.

Pachauri, R.K., Meyer, L.A., et al., (Eds.), 2014. IPCC, 2014: Climate Change 2014: Synthesis Report. Contribution of Working Groups I, II and III to the Fifth Assessment Report of the Intergovernmental Panel on Climate Change IPCC, Geneva, Switzerland, 151 pp.

Pahuja, N., Pandey, N., Mandal, K., Bandopadhay, C., 2014. GHG Mitigation in India: An Overview of Current Policy Landscape. World Resource Institute (WRI), Washington, DC, Working paper, pp. 32.

Pakistan INDC, 2016. Pakistan Intended Nationally Determined Contributions Report, pp. 31.

Pant, G.B., 2003. Long term climate variability and change over monsoonal area. J. Indian Geophys. Union 7 (3), 125–134.

Patz, J.A., Campbell-Lendrum, D., Holloway, T., Foley, J.A., 2005. Impact of regional climate change on human health. Nature 17, 310–317.

Paul, B., Rashid, H., 1993. Flood damage to rice crop in Bangladesh. Geogr. Rev. 83 (2), 151–159.

Pelling, M., 2011. Adaptation to Climate change – From resilience to Transformation. Routledge Taylor and Fransis Group, 2 Park Square, Milton park, Abingdon, oxon, OX14 4RN, UK, pp. 1–274.

Percival, L.M.E., Witt, M.L.I., Mather, T.A., Hermoso, M., Jenkyns, H.C., Hesselbo, S.P., Al-Suwaidi, A.H., Storm, M.S., Xu, W., Ruhl, M., 2015. Globally enhance mercury deposition during the end-Pleinsb-achian extinction and Toarcian OAE: a link to the Karroo-Ferrar large igneous province. Earth Planet. Sci. Lett. 428, 267–280.

Perkins, P.E., 2014. International Partnerships of women for Sustainable Watershed Governance in Times of Climate Change. Paper submitted for inclusion in the book, A Political Ecology of Women, Water and Global Environmental Change edited by Stephanie Buechler and Anne-Marie Hanson.

Pernetta, J.C. (Ed.), 1993. Marine Protected Area Needs in the South Asian Seas Region: Development of a System. A Marine Conservation and Development Report. IUCN, Gland, Switzerland, pp. 1–72.

Petit, J.R., Jougel, J., Raynaud, D., Barkov, N.I., Barnola, J.M., Basile, I., Bender, M., Chappelaz, J., Davis, M., Delaygue, G., Delmotte, M., Kotlyakov, V.M., Legrand, M., Lipenkov, V.Y., Lorius, C., Pepin, L., Ritz, C., Saltzman, E., Stievenard, M., 1999. Climate and atmospheric history of the past 420,000 years from the Vostok ice core Antarctica. Nature 399, 429–436.

Petersen, I., Maarais, D., Abdulmalik, J., Ahuja, S., et al., 2017. Strengthening mental health system governance in six low and middle income countries in Africa and South Asia: challenges, needs and potential strategies. Health Policy Plan 32, 699–709.

Pielke, R.A. JJr., 1998. Rethinking the role of adaptation in climate policy. Glob. Environ. Change 8, 159–170.

Piran, K., Dedekorkut-Howes, A., 2015. Noosa climate action plan: an example of collaborative climate governance? State of Australian Cities Conference. pp. 1–16.

Prell, C., 2015. Paris Agreement on Climate Change: A Practical Guide. Crowell & Moring.

Rahman, S.H., Wijayatunga, P.D.C., Gunatilake, H., Fernando, P.N., 2012. Energy Trade in South Asia: Opportunities and Challenges. Asian Development Bank, Manila. Available from: http://www.adb.org/sites/default/files/publication/29703/energy-trae-south-asia.pdf.

Rajan, P., Gangbar, J., Gayathri, K., 2014. Child and mental health and nutrition in South Asia-lessons for India. Working paper 323, Centre for Economic Studies and Policy, The Institute for Social and Economic Change, Bangalore.

Raju, K.V., 2006. Bio-fuels in south Asia: an overview. Asian Biotechnol. Dev. Rev. 8 (2), 1–9.

Ramachandra, T.V., Shwetmala, 2012. Decentralized carbon footprint analysis for opting climate change mitigation strategies in India. Renew. Sustain. Energy Rev. 16, 5820–5833.

Ramanathan, V., Carmichael, G., 2008. Global and regional climate changes due to black carbon. Nat. Geosci. 1, 221–227.

Ramanathan, V., Chung, C., Kim, D., Bettge, T., Buja, L., Kiehl, J.T., Washington, W.M., Fu, Q., Sikka, D.R., Wild, M., 2005. Atmospheric Brown Clouds: Impacts on South Asian Climate and Hydrological Cycle Proceedings of National Academy of Sciences. vol. 102 (15), pp. 5326–5333.

Ramanathan, V., Feng, Y., 2009. Air pollution, greenhouse gases and climate change: global and regional perspectives. Atmos. Environ. 43, 37–50.

Rana, P.B., Chia, W.M., 2015. Economic preforms in South Asia: an overview and the remaining agenda. In: Working paper. S. Rajaratnam School of International Studies, Singapore, No. 289.

Ranasinghe, D.M.K.H., 2010. Climate change mitigation-Sri Lanka's perspective. In: Proceedings of the 15th International Forestry and Environment Symposium, 26–27 Nov 2010. Department of Forestry and Environmental Science, University of Sri Jayewardenapura, Sri Lanka, pp. 290–296.

Rappaport, E.N., 2014. Fatalities in the United States from Atlantic tropical cyclones: new data and interpretation. Bull. Am. Meteorol. Soc. 95, 341–346.

Rasul, G., 2014. Food, water, and energy security in South Asia: a nexus perspective from the Hindu Kush Himalayan region. Environ. Sci. Policy 39, 35–48.

Raut, A., 2006. Climate impacts on Nepal. Tiempo 60, 3–5.

Ravindranath, N.H., Chaturvedi, R.K., Kumar, P., 2017. Paris Agreement; research, monitoring and reporting requirements for India. Curr. Sci. 112 (5), 916–922.

Ray, S., 2008. Comparative study of virgin and reclaimed islands of Sundarban mangrove ecosystem through network analysis. Ecol. Modell. 215, 207–216.

Resio, D.T., Westerink, J.J., 2008. Modeling the physics of storm surges. Phys. Today 61, 33–38.

Rignot, E., Velicogna, I., Van den broeke, M.R., Monaghan, A., Lenaerts, J.T.M., 2011. Acceleration of the contribution of the Greeland and Antarctic ice-sheets to sea level rise. Geophys. Res. Lett. 38, 1–5, L05503.

Ritchie, H., Roser, M., 2017. CO2 and other greenhouse gas emissions. Available from: https://ourworldindata.org.

Robock, A., 1994. Review of year without a summer? World climate in 1816. Clim. Change 26, 105–108.

Rodbell, D.T., Seltzer, G.O., Anderson, D.M., Abbott, M.B., Enfield, D.B., Newman, J.H., 1999. A similar to 15000 year record of El Nino driven alleviation in south-western Ecuador. Science 283, 516–520.

Rodrik, D., 2013. The past, present and future of the economic growth. Working Paper 1, June 2013. Global Citizen Foundation, pp. 1–58.

Rogelj, J., Meinshausen, M., Sedlácek, J., Knutti, R., 2014. Implications of potentially lower climate sensitivity on climate projections and policy. Environ. Res. Lett. 9 (3), 1–7, 031003.

Rojas-Downing, M.M., Nejadhashemi, A.P., Harrigan, T., Woznicki, S.A., 2017. Climate change and livestock: Impacts, adaptation, and mitigation. Clim. Risk Manag. 16, 145–163.

Roy, A., 2010. Vulnerability of the Sundarbans ecosystem. J. Coast. Environ. 1, 169–181.

Royal Society, 2009. Royal Society "Geoengineering the Climate: Science, Governance and Uncertainty." Geoengineering the Climate: Science, Governance and Uncertainty (Report). London, UK. p. 1 (September, 2009).

Russell, S., 2013. Nitrogen Trifluoride Now Required in GHG Protocol Greenhouse Gas Emission Inventories. World Resource Institute Blog.

SAARC Portal, 2016. South Asian Association for Regional Cooperation Web Portal 2016.

SAARC, 2014. Eighteenth SAARC Summit Declaration, Khatmandu, Nepal.

Sarkar, S.K., Bhattacharya, B., Debnath, S., Bandopadhyay, G., Giri, S., 2002. Heavy metals in biota from Sundarbans Wetland Ecosystem, India: implications to monitoring and environmental assessment. Aquat. Ecosys. Health Manage. 5, 467–472.

Satpathy, B., Muniapan, B., Dass, M., 2013. UNESCAP's characteristics of good governance from the.philosophy of Bhagavad-Gita and its contemporary relevance in the Indian context. Int. J. Indian Cult. Bus. Manage. 7 (2), 192–212.

Scheffer, M., Carpenter, S., Foley, J.A., Folke, C., Walker, B., 2001. Catastrophic shifts in ecosystems. Nature 413, 591–596.

Schmidt, G.A., Ruedy, R.A., Miller, R.L., Lacis, A.A., 2010. Attribution of the present-day total greenhouse effect. J. Geophys. Res. 115 (1-6).

Schneider, S.H., Semenov, S., Patwardhan, A., Burton, I., Magadza, C.H.D., Oppenheimer, M., Pittock, A.B., Rahman, A., Smith, J.B., Suarez, A., Yamin, F., 2007. Executive summary, Chapter 19: Assessing key vulnerabilities and the risk from climate change. In: Parry, M.L., Canziani, O.F., Palutikof, J.P., van der Linden, P.J., Hanson, C.E. (Eds.),Climate Change 2007: Impacts, Adaptation and Vulnerability. Contribution of Working Group II to the Fourth Assessment Report of the Intergovernmental Panel on Climate Change. Cambridge University Press, Cambridge, UK.

Scotese, C.R., 2011. Cretaceous Paleogeography and plate tectonic reconstructions. The PALEOMAP Project PaleoAtlas for Arc-GIS, 2, Arlington, TX.

Scotese, C.R., 2015. Some Thoughts on Global Climate Change: the Transition From Icehouse to Hothouse. Paleomap project, vol. 21a, pp. 1–55.

Self, S., Schmidt, A., Mather, T.A., 2014. Emplacement characteristics, time scales, and volcanic gas release rates of continental flood basalt eruptions on earth. In: Keller, G., Kerr, A.C. (Eds.), Volcanism, Impacts, and Mass Extinctions; Causes and Effects. Geological Society of America Special Paper 2014; 505, 319-337.

Sethi, N., 2007. Global warming: Mumbai to face the heat. Times of India. March 18, 2007.

Shadid, W.A., 2007. Grondslagen van de interculturele communicatie. Studieveld en werkterrein. Bohn Stafleu van Loghum: Houten/Diegem. (Also The connection between culture and climate change by Bernadet van den Pol).

Shah, T., 2007. The ground water economy of south Asia: an assessment of size significance and socio-ecological impacts. In: Giordano, M., Villholth, K.G. (Eds.), CAB International 2007. The Agricultural Groundwater Revolution: Opportunities and Threats to Development, pp. 1–30 (Chapter 2).

Shamsudduha, M., Uddin, A., 2007. Quaternary shoreline shifting and hydrogeologic influence on the distribution of groundwater arsenic in aquifers of the Bengal Basin. J. Asian Earth Sci. 31, 177–194.

Sharma, G., 2015. Hydropower development in India: untapped and unchartered territory. Power Watch India, 1-5. Available from: powerwatchindia.com.

Shepherd, J.G., 2009. Geoengineering the climate: science, governance and uncertainty. Royal Society report, Sept 2009. Available from: http://eprints.soton.ac.uk/156641/1//Geoengineering_the_climate.pdf.

Shindell, D., Kuylenstierna, J.C.I., Vignati, E., van Dingenen, R., Amann, M., Klimont, Z., Anenberg, S.C., Muller, N., Janssens- Maenhout, G., Raes, F., Schwartz, J., Faluvegi, G., Pozzoli, L., Kupiainen, K., Hoglund-Isaksson, L., Emberson, L., Streets, D., Ramanathan, V., Hicks, K., Oanh, N.T.K., Milly, G., Williams, M., Demkine, V., Fowler, D., 2012. Simultaneously mitigating near-term climate change and improving human health and food security. Science 335, 183–189.

Shrestha, A.B., Wake, C.P., Mayewski, P.A., Dibb, J.E., 1999. Maximum temperature trends in the Himalaya and its vicinity: an analysis based on temperature records from Nepal for the period 1971–94. J. Clim. 12, 2775–2786.

Shrestha, R.M., Ahmed, M., Suprachalasi, G., Lasco, R., 2012. Economics of Reducing Greenhouse Gas Emissions in South Asia: Options and Cost, ADB-Australian AID Report, p. 159.

Shriver, D., Atkins, P., 2010. Inorganic Chemistry. In: Freeman, W.H (Ed.). p. 409.

Shukla, A.K., Sudhakar, K., Baredar, P., 2017. Renewable energy resources in south Asian countries: challenges, policies and recommendations. Resour. Efficient Technol., 1–5.

Shukla, P.K., Garg, A., Ghosh, D., 2001. Future energy trends and GHG emissions for India, climate change economics and policy, Indian perspectives. In: Toman, M. (Ed.), Resources for the Future Publications, Washington, DC, Available from: www.ezanalytics.com.

Sikdar, C., Mukhopadhyay, K., 2016. Impact of population on carbon emission: Lessons from India. Asia Pac. Dev. J. 23 (1), 105–132.

Sikri, V., 2015. Gender, community and violence: changing mindsets for empowering the women of South Asia 15-16 April, 2015. In: Conference on Gender, Community and Violence: Changing Mindsets for Empowering the Women of South Asia. SWAN, Jamia Millia Islamia University, New Delhi.

Silvia, C., 2011. Collaborative concepts for successful network leadership. State Local Govern. Rev. 43, 66–71.

Singh, H.K., Chandarsekharam, D., Trupti, G., Mohite, P., Singh, B., Varun, C., Sinha, S.K., 2016. Potential geothermal energy resource of India: a review. Curr. Sustain./Renew. Energy Rep. 3 (3), 80–91.

Sivakumar, M.V.K., Stefanski, R., 2011. In: Lal, R., Sivakumar, M.V.K., Faiz, M.A., Mustafizur Rehman, A.H.F., Islam, K.R. (Eds.), Climate Change and Food Security in South Asia. Springer Science, Business Media B.V, p. 600.

Sloss, L., 2012. Black carbon emissions in India CCP/209. IEA Clean Coal Centre. pp. 1–38.

Solomon, S., Karen, H.R., Daniel, J.S., Davis, S.M., Sanford, T.J., Plattner, G.K., 2010. Contributions of stratospheric water vapor to decadal changes in the rate of global warming. Science 327, 1219–1223.

Somaratne, S., Dhanapala, A.H., 1996. Potential impact of global climate change on forest in Sri Lanka. In: Erda, L., Bolhofer, W., Huq, S., Lenhart, S., Mukherjee, S.K., Smith, J.B., Wisniewski, J. (Eds.), Climate Change Variability and Adaptation in Asia and the Pacific. Kluwer, Dordrecht, The Netherlands, pp. 129–135.

Somerville, R., 2008. The Ethics of Climate Change. Available from: www.e360.yale.edu.

Sousounis, K., 2016. How is Climate Change a Feminist Issue? Available from: www.girlup.org.

Sridhar, K.S., 2010. Carbon emissions, climate change and impacts in India's cities, in India infrastructure report 2010. In: 3iNetwork (Ed), Infrastructure Development in a Sustainable Low Carbon Economy: Road Ahead for India. Oxford University Press, New Delhi, pp. 345–354.

Srivastava, R., 2016. Tidal energy: an overview. Renewable Technologies in Nascent Stages: Tidal Power, Uses and Current Tidal Power Plants in India. Green World Investor. vol. 4.

Sterrett, C., 2011. Review of Climate Change Adaptation Practices in South Asia. Oxfam Research Reports, Oxfam GB, Oxford, UK.

Strauss, S., 2012. Are cultures endangered by climate change? Yes, but… WIREs. Clim. Change. 181, pp. 1–6. doi:10 1002/wcc.

Sultana, H., Ali, N., 2006. Vulnerability of wheat production in different climatic zones of Pakistan under climate change scenarios using CSM-CERES-Wheat Model. Second International Young Scientists' Global Change Conference. 7–9 November 2006, Beijing.

Sutton, R.W., Dong, B., Gregory, J.M., 2007. Land/sea warming ratio in response to climate change: IPCC AR4 model results and comparison with observations. Geophys. Res. Lett. 34 (1-5), L02701.

Tanner, T., Mitchell, T., Polack, E., Guenther, B., 2009. Urban Governance for Adaptation: Assessing Climate Change Resilience in Ten Asian Cities. Institute of development studies, UK, Working paper 315, 47.

Technology Needs Assessment (TNA) Report, 2016. Climate Change Mitigation. Govt. of Pakistan, Ministry of Climate change, Islamabad, Pakistan. Funded by Global Environment Facility (GEF) and implemented by UNEP and UNEP DTU Partnership in collaboration with Regional Center, Asian Institute of Technology, Bangkok, pp. 1–112.

The Energy Resources, Institute (TERI), 2015. Green growth and climate change mitigation, New Delhi. p. 36.

Thapa, G., 2004. Rural Poverty Reduction Strategy for South Asia. In: ASARC Working Paper 2004–06, Paper Presented at an International Conference on Ten Years of Australian South Asia Research Centre organized at the Australian National University, Canberra from 27–28 April 2004.

The Guardian, 2016. Renewables made up half of net electricity capacity added last year. Available from: https://www.theguardian.com/environment/2016/oct/25/renewables-made-up-half-of-net-electricity-capacity-added-last-year.

Tiwari, K.R., Rayamajhi, S., Pokharel, R.K., Balla, M.K., 2014. Does Nepal's climate change adaptation policy and practices address poor and vulnerable communities?. J. Law Policy Glob. 23, 28–38.

TOI, 2015. Livestock causes 15% of all emissions worldwide. Times of India, 16 December 2015.

Tomkins, F., Deconscini, C., 2015. 2014: A year of temperature records and landmark climate findings. World Resource Institute (WRI), Fact Sheet, pp. 1–4.

Topriska, E., 2016. Hydrogen could become the new fuel for cooking—here's how. The Conversation Media Group, Australia, pp. 1–7.

Transparency International, 2016. Corruption Perceptions Index, pp. 1-9. Available from: www.transparency.org.

Tsitsiragos, D., 2016. Climate change is a threat – and an opportunity – for the private sector. Opin. Capital Finance Int. Mag. 1–4.

Tubiello, F., 2012. Climate change adaptation and mitigation. Challenges and Opportunities in the food sector. Food and Agriculture Organization of the United Nations (FAO). Natural Resources management and Environment Department, Rome. 2012. Prepared for the High-level conference on world food security: the challenges of climate change and bioenergy. Rome, 3-5 June, 2008.

Tyndall, J., 1861. The Bakerian lecture: on the absorption and radiation of heat by gases and vapours, and on the physical connexion of radiation, absoprtion, and conduction. Philos. Trans. Royal Soc. Lond. 151, 1–36.

UNDP, 2013. Human Development Report, The Rise of South: Human Progress in Diverse World. United Nation Development Programme, New York, 202.

UNEP-GRID-Arendal, 2012. Climate change mitigation in India, pp 1–34, Norway. Available from: www.grida.no.

United Nations Economic and Social Commission for Asia and the Pacific (UNESCAP), 2012. Green Growth resources and Resilience, Environmental Sustainability in Asia and the Pacific. United Nations and Asian Development Bank, Bangkok, pp. 1–157.

UNESCAP, 2016. Achieving the sustainable development goals in South Asia: key policy priorities and implementation challenges. United Nations Economic and Social Commission for Asia and the Pacific, pp. 1–31.

UNESCAP (United Nations Economic and Social Commission for Asia and the Pacific) Economic and Social survey of Asia and the Pacific 2017 Governance and Fiscal Management. United Nations publication Sales No. E.17.II.F.8, United Nations 2017, Printed in Bangkok, ISBN: 978-92-1-120742-2.

UNEPDA (United Nations Environment Programme and Development Alternatives). South Asia Environment Outlook : UNEP, SAARC and DA. UNEP Regional office for Asia and Pacific (UNEP/ROAP), Bangkok, 10200, pp. 302, 2014.

UNFCCC, 2015. UN Climate Change Centre, Bonn Climate Change Conference, United Nations Framework Convention on Climate Change.

UNFCCC, 2017. United Nations Framework Convention on Climate Change. In: Bonn Climate Change Conference. UN Climate Change Centre, May 2017.

Unnikrishnan, A.S., Shankar, D., 2007. Are sea level rise trends along the coasts of the north Indian Ocean consistent with global estimates? Glob. Planet. Change 57, 301–307.

US AID 2015. United States AID Agency Financial Report, p. 168.

UN.org via Wikimedia Commons

USEPA, 2016. United States Environmental Protection Agency. Climate Change Indicators in the United States: Atmospheric Concentrations of Greenhouse Gases. pp. 1-13. Available from: www.epa.gov/climate-indicators.

US Environmental Protection Agency 2010. Washington, DC, USA.

US Global Change Research Program (USGCRP), 2014. Our Changing Planet, Indicators of Change, Adapting to Change. USGCRP Released the Third National Climate Assessment, the Authoritative and Comprehensive Report on Climate Change and its Impacts in the United States. Available from: http://www.globalchange.gov/.

Unites States Greenhouse Gas Inventory Report (USGHGIR), 2011. Climate Change Division Office of Atmospheric Programs. US Environmental Protection Agency, Pennsylvania Ave NW, Washington, DC, (MC-6202A) 20460 1200.

van Andel, N., 2011. Climate Changes are Not Caused by Greenhouse Gases. Available from: http://hockeyschtick.blogspot.com/2011/01/scientist-climate-changes-are-not.html.

Verweij, M., Thompson, M. (Eds.), 2006. Clumsy Solutions for a Complex World. Macmillan, London, Palgrave.

Visweswaran, K., 2004. Gendered states: rethinking culture as a site of South Asian Human Rights Work. Hum. Rights Q. 26, 483–511.

Walker, M., Johnson, S., Rasmussen, S.O., et al., 2009. Formal definition and dating of the GSSP (global stratotype section and point) for the base of the Holocene using the Greenland NGRIP ice cores and selected auxiliary records. J. Quatern. Res. 24, 3–17.

Walsh, J., Wuebbles, D., Hayhoe, K., Kossin, J., Kunkel, K., Stephens, G., Thorne, P., Vose, R., Wehner, M., Willis, J., Anderson, D., Doney, S., Feely, R., Hennon, P., Kharin, V., Knutson, T., Landerer, F., Lenton, T., Kennedy, J., Somerville, R., 2014a. Our changing climate, climate change impacts in the United States. In: Melilo, J.M., Richmond, T.C., Yohe, G.W. (Eds.), The Third National Climate Assessment. US Global Change Research Program, pp. 19–67. (Chapter 2). doi:107930/JOKW5CXT.

Walsh, J., Wuebbles, D., Hayhoe, K., Kossin, J., Kunkel, K., Stephens, G., Thorne, P., Vose, R., Wehner, M., Willis, J., Anderson, D., Kharin, V., Knutson, T., Landerer, F., Lenton, T., Kennedy, J., Somerville, R., 2014b. Appendix 3: climate science supplement, climate change impacts in the United States. In: Melilo, J.M., Richmond, T.C., Yohe, G.W. (Eds.), The Third National Climate Assessment. US Global Change Research Program, pp. 735–789. doi:107930/JOKS6PHH.

Warrick, R.A., Le Provost, C., Meier, M.F., Oberlemans, J., Woodworth, P.L., 1996. Changes in sea level. In: Houghton, J.T., Filho, L.T.M., Calender, B.A., Harris, N., Kattenberg, A., Maskel, K. (Eds.), Climate Change 1995: The Science of Climate Change IPCC, WMO, UNEP. Cambridge University Press, Cambridge, p. 371.

WEC, 2013. World Energy Resources: Marine Energy. World Energy Council, UK, pp. 444–455 (Chapter 11).

WER—World Energy Resources, 2016. Executive Summary, World Energy Council, London, UK, 6p.

Wijayatunga, P., Fernando, P.N., 2013. An Overview Of Energy Co-Operation In South Asia. No.19, May 2013, Asian Development Bank South Asia Working Paper Series. Asian Development Bank, 6 ADB Avenue Mandaluyong City, 1550 Metro, Manila, Philippines, pp. 1–43.

Wijayatunga, P., Simbalapitiya, T., 2016. Improving Energy Efficiency in South Asia. No. 47, ADB South Asia Working Paper Series. ADB, Mandaluyong City, Philippines, pp. 1–29.

Wijeratne, M.A., 1996. Tea: plucking strategies. Planters chronicle: September 1996, p. 443.

Williamson, M., 2017. Disaster Risk in South and South West Asia: A Pathway From Risk to Resilience. United Nations Economic and Social Commission for Asia and the Pacific (UN-ESCAP) Report. South and South-West Asia office, Kathmandu, Nepal, pp. 1–19.

Wilson, J., 2011. Water Security In South Asia: Issues and Policy Recommendations.ORF (Observer Research Foundation) Issue Brief, #26, 1-11, 2011. Observer Research Foundation, 20, Rouse Avenue, New Delhi-110 002, India.

Wingham, D.J., Siegert, M.J., Shepherd, A., Muir, A.S., 2006. Rapid discharge connects Antarctic subglacial lakes. Nature 440, 1033–1036.

Wolpert, S., 1999. A New History of India, sixth ed. Oxford University Press, ISBN 978-0-19-512877-2.

World Bank 2012a. South Asia Region Disaster Risk Management and Climate Change Unit: A Regional Overview, 1-116

World Bank, 2012b. World Bank development Report. Gender Equality and Development. The International Bank for Reconstruction and Development/The World Bank, Washington DC, USA, 1–458.

World Bank, 2014. Annual Report. Copy right. 2014 International Bank for Reconstruction and Development. The World Bank, Washington, DC.

World Bank, 2015. The World Bank Annual Report 2015. Washington, DC. Available from: https://openknowledge.worldbank.org/handle/10986/22550.

World Bank, 2017. Ease of doing Business. Available from: https://data.worldbank.org/indicator//.

World Resources Institute (WRI), 2015. Washington, DC, USA, p. 64.

World Resources Institute (WRI), 1996. World Resources: 1994–95: A Guide to the Global Environment. Oxford University Press.

Xiao, G.W., Liu, Q., Xu, Z., Sun, Wang, J., 2005. Effects of temperature increase and elevated CO_2 concentration, with supplemental irrigation, on the yield of rain-fed spring wheat in a semi-arid region of China. Agric. Water Manage. 74, 243–255.

Yadav, H.J., 2015. Recent developments of ocean energy in India-an overview. Int. J. Recent Innov. Trends Comput. Commun. 3 (12), 6717–6721.

Yang, B., Liu, C., Su, Y., Jing, X., 2017. The allocation of carbon intensity reduction target by 2020 among industrial sectors in China. Sustainability 9, 148–166.

Zacharias, N., 2008. Anthropogenic and Natural Causes of Climate Change: Physical Fundamentals of Global Change Process Lecture. Fachhochschule Eberswalde, Germany, p. 18.

Zeigler, P.A., 2000. Use of paleorecords in determining variability within the volcanism-climate system. Quat. Sci. Rev. 19, 417–438.

Zeigler, P.A., 2013. Climate Change During Geological and Recent Times, Anything New or déjà vu? Causes, Speculations and IPCC Postulates. Available from: www.friendsofscience.org.

Zielinski, G.A., 2000. Use of paleorecords in determining variability within the volcanism-climate system. Quat. Sci. Rev. 19, 417–438.

Index

Printed in the United States
By Bookmasters